Computer Science

for Cambridge IGCSE™ & O Level

**PROGRAMMING BOOK
FOR MICROSOFT® VISUAL BASIC**

Richard Morgan

Shaftesbury Road, Cambridge CB2 8EA, United Kingdom

One Liberty Plaza, 20th Floor, New York, NY 10006, USA

477 Williamstown Road, Port Melbourne, VIC 3207, Australia

314–321, 3rd Floor, Plot 3, Splendor Forum, Jasola District Centre, New Delhi – 110025, India

103 Penang Road, #05–06/07, Visioncrest Commercial, Singapore 238467

Cambridge University Press & Assessment is a department of the University of Cambridge.

We share the University's mission to contribute to society through the pursuit of education, learning and research at the highest international levels of excellence.

www.cambridge.org
Information on this title: www.cambridge.org/9781108935678

© Cambridge University Press & Assessment 2021

This publication is in copyright. Subject to statutory exception and to the provisions of relevant collective licensing agreements, no reproduction of any part may take place without the written permission of Cambridge University Press & Assessment.

First published 2015
Second edition 2021

20 19 18 17 16 15 14 13 12 11 10 9 8 7 6 5 4 3 2

Printed in Great Britain by CPI Group (UK) Ltd, Croydon CR0 4YY

A catalogue record for this publication is available from the British Library

ISBN 978-1-108-93567-8 Programming Book Paperback with Digital Access (2 Years)
ISBN 978-1-108-94084-9 Digital Programming Book (2 Years)

Additional resources for this publication at www.cambridge.org/go

Cambridge University Press & Assessment has no responsibility for the persistence or accuracy of URLs for external or third-party internet websites referred to in this publication, and does not guarantee that any content on such websites is, or will remain, accurate or appropriate. Information regarding prices, travel timetables, and other factual information given in this work is correct at the time of first printing but Cambridge University Press & Assessment does not guarantee the accuracy of such information thereafter.

Exam-style questions and sample answers have been written by the authors. In examinations, the way marks are awarded may be different. References to assessment and/or assessment preparation are the publisher's interpretation of the syllabus requirements and may not fully reflect the approach of Cambridge Assessment International Education.

The information in Chapter 12 is based on the Cambridge IGCSE, IGCSE (9–1) and O Level Computer Science syllabuses (0478/0984/2210) for examination from 2023. You should always refer to the appropriate syllabus document for the year of your examination to confirm the details and for more information. The syllabus documents are available on the Cambridge International website at *www.cambridgeinternational.org*

NOTICE TO TEACHERS IN THE UK
It is illegal to reproduce any part of this work in material form (including photocopying and electronic storage) except under the following circumstances:
(i) where you are abiding by a licence granted to your school or institution by the Copyright Licensing Agency;
(ii) where no such licence exists, or where you wish to exceed the terms of a licence, and you have gained the written permission of Cambridge University Press & Assessment;
(iii) where you are allowed to reproduce without permission under the provisions of Chapter 3 of the Copyright, Designs and Patents Act 1988, which covers, for example, the reproduction of short passages within certain types of educational anthology and reproduction for the purposes of setting examination questions.

DEDICATED TEACHER AWARDS

Teachers play an important part in shaping futures. Our Dedicated Teacher Awards recognise the hard work that teachers put in every day.

Thank you to everyone who nominated this year; we have been inspired and moved by all of your stories. Well done to all of our nominees for your dedication to learning and for inspiring the next generation of thinkers, leaders and innovators.

Congratulations to our incredible winner and finalists!

WINNER

- **Patricia Abril** — New Cambridge School, Colombia
- **Stanley Manaay** — Salvacion National High School, Philippines
- **Tiffany Cavanagh** — Trident College Solwezi, Zambia
- **Helen Comerford** — Lumen Christi Catholic College, Australia
- **John Nicko Coyoca** — University of San Jose-Recoletos, Philippines
- **Meera Rangarajan** — RBK International Academy, India

For more information about our dedicated teachers and their stories, go to
dedicatedteacher.cambridge.org

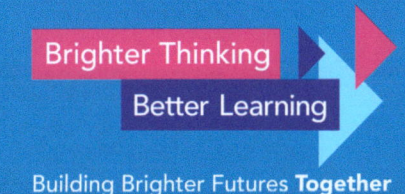

Brighter Thinking
Better Learning

Building Brighter Futures Together

Contents

The items in orange are available on the digital edition that accompanies this book.

Introduction	vi
How to use this book	viii
How to use this series	x

1 Visual Studio Community

1.1	Getting Visual Studio 2019 Community for Windows	2
1.2	The Integrated Development Environment	3
1.3	Getting started	4
1.4	Console application	5
1.5	Make your first program using console application	7

Note: Windows Forms application is optional content. It is not covered in the syllabuses.

1.6	Windows Forms application	10
1.7	Make your first Windows Forms application	13
1.8	Choosing a console application or Windows Forms application	19
1.9	Additional support	20

2 Programming constructs

2.1	Programming concepts	22
2.2	Design tools	24
2.3	Flowcharts	24
2.4	Pseudocode	34
2.5	Use of flowcharts and pseudocode in programming	35

3 Variables and arithmetic operators

3.1	Variables and constants	39
3.2	Types of data	39
3.3	Pseudo numbers	40
3.4	Declaring variables and constants	41
3.5	Variable scope	42
3.6	Arithmetic operators	43
3.7	More complex arithmetic operators	43
3.8	Programming	45
3.9	Commenting code	48
3.10	Making real-world scenario-based systems	49
3.11	Rounding function	59
3.12	Random function	60

4 Selection

4.1	The need for selection	64
4.2	IF statements	64
4.3	Single action IF statements	65
4.4	Logical operators	65
4.5	Coding IF statements in Visual Basic	66
4.6	Multiple decisions	68
4.7	CASE statements	74
4.8	Coding CASE statements in Visual Basic	75
4.9	Creating complex conditions	82
4.10	Using Boolean operators to connect criteria	83
4.11	Connecting more than one value	86

5 Iteration

5.1	Types of iteration	93
5.2	FOR loops (count-controlled loops)	93
5.3	Using the loop counter	95
5.4	Combining iteration and selection	99
5.5	Common iterative tasks	102
5.6	Condition-controlled loops	107
5.7	WHILE loops (pre-condition loops)	107
5.8	WHILE loops with multiple criteria	113
5.9	REPEAT...UNTIL loops (post-condition loops)	116
5.10	WHILE and REPEAT...UNTIL loops based on user input	119
5.11	Choosing to use WHILE or UNTIL	123
5.12	Nested iteration	123

6 Designing algorithms

6.1	Top-down design	129
6.2	Structure diagrams	129
6.3	Design steps	132
6.4	The complete design process	138

7 Subroutines and file handling

7.1	Why use subroutines?	141
7.2	Subroutines in Visual Basic	142
7.3	Functions	143
7.4	Procedures	149
7.5	File handling	155

8 Checking inputs

8.1	Validation	163
8.2	Verification	164
8.3	Programming validation into your systems	165

9 Testing

9.1	When to test	186
9.2	Debugging	186
9.3	IDE debugging tools and diagnostics	188
9.4	Identifying logical errors	190
9.5	Dry running	190
9.6	Breakpoints, variable tracing and stepping through code	193
9.7	Beta testing	195

10 Arrays

10.1	What is an array?	201
10.2	Declaring a one-dimensional array	201
10.3	Initialising arrays	202
10.4	Using arrays	203
10.5	Iteration in arrays	208
10.6	Grouped data records	212
10.7	Two-dimensional arrays	217
10.8	Sorting	224
10.9	Sorting algorithms	225
10.10	Bubble sort	226

11 Manipulating strings

11.1	Concatenation	235
11.2	Splitting strings	237
11.3	String index	237
11.4	Substring	241
11.5	Looping through all the characters in a string	244

12 Programming scenario task

12.1	Getting started on the question	248
12.2	Planning the program	249
12.3	Planning your solution	250
12.4	Writing your program	251
12.5	Checking your program	253
12.6	Final thoughts	254

13 Examination practice — 258

Glossary — 268

Acknowledgements

Solutions

> Introduction

I have written this book with two aims in mind. The first is to provide a programming book that specifically covers the material relevant to the Cambridge IGCSE™, IGCSE (9–1) and O Level Computer Science syllabuses (0478/0984/2210). The second, and perhaps more important, aim is to provide you with a start to the exciting and rewarding process of being able to create your own computer programs.

Language

The syntax and structures used to implement programming techniques will vary across different languages. The book is entirely based around Visual Basic, one of the three recommended languages for the syllabuses. Similar books are also available which focus on the Python and Java programming languages.

Microsoft® Visual Basic offers you, as a programmer, two modes of application. There is a simple console window in which you can learn and develop programming skills. It also offers a Windows Forms application, which allows you to program commercial-style applications that offer a graphical user interface through which users can interact with programs. The language is supported by a fully functional development environment called Visual Studio Community which is available free directly from Microsoft. They also provide excellent support and language-specific tutorials via the Microsoft Developer Network.

Support

As you work your way through the exercises in this book you will develop your computational skills, independent of any specific programming language. You will do this through the use of program design tools such as structure diagrams and flowcharts. You will also make use of pseudocode, a structured method for describing the logic of computer programs.

It is crucial that you become familiar with these techniques. Throughout this book all the programming techniques are demonstrated in the non-language-specific format. To support learning and assessment, all the chapters include end-of-chapter tasks. Solutions to all the end-of-chapter tasks can be found on the accompanying digital version of this book. Finally, Chapter 13 has exam-style questions and a sample mark scheme prepared by the author, giving possible solutions.

Developing programming skills

One of the advantages of Visual Basic is that it provides a language that encourages you to program solutions making use of the basic programming constructs: sequence, selection and iteration. Although the language does have access to many powerful prewritten code libraries, they are not generally used in this book.

Computational thinking is the ability to resolve a problem into its constituent parts and to provide a logical and efficient coded solution. Experience of teaching GCSE and A level computer science for more than 15 years tells me that knowing how to think computationally relies much more on an understanding of the underlying programming concepts than on the ability to learn a few shortcut library routines. This book is aimed at teaching those underlying skills which can be applied to the languages of the future.

It is without doubt that programming languages will develop over the coming years but the ability to think computationally will remain a constant. As technology increasingly impacts on society, people with computation thinking skills will be able to help shape the way that technology impacts on our future.

CAMBRIDGE IGCSE™ & O LEVEL COMPUTER SCIENCE: PROGRAMMING BOOK

> How to use this book

Throughout this book, you will notice lots of different features that will help your learning. These are explained below.

LEARNING INTENTIONS

These set the scene for each chapter, help with navigation through the Visual Basic programming process and indicate the important concepts in each topic.

SKILLS FOCUS

This feature supports your computational thinking, mathematical and programming skills. They include useful explanations, step-by-step examples and questions for you to try out yourselves.

Pseudocode and Code snippets

This structured method for describing the logic of computer programs may be very similar to the pseudocode used in the syllabuses.

```
Dim Number As Integer
Number = Console.ReadLine()

If Number > 0 Then
    If Number Mod 2 = 0 Then
        'accept number'
    Else
        'reject number'
    End If
Else
    'reject number
End If
```

Code snippet 4.6: Different coded approaches to the positive even number task

Pseudocode is shown in the text like this:

```
// Entering the values
INPUT Number1
INPUT Number2
// Calculate the addition and store in Answer
Answer ← Number1 + Number2
// Output the value in Answer
OUTPUT Answer
```

Further Information: This feature highlights the advanced aspects in this book that go beyond the immediate scope of the syllabuses.

KEY WORDS

Key vocabulary is highlighted in the text when it is first introduced. Definitions are then given in the margin, which explain the meanings of these words and phrases. You will also find definitions of these words in the glossary at the back of this book.

TIP

These are short suggestions to remind you about important learning points. For example, a tip to help clear up misunderstandings between pseudocode and Visual Basic.

How to use this book

Programming tasks

Programming tasks give you the opportunity to develop your programming and problem-solving skills. Answers to these questions can be found in the solutions chapter, on the digital part of this resource. There are three different types of programming tasks:

DEMO TASKS

You will be presented with a task and a step-by-step solution will be provided to help familiarise you with the techniques required.

PRACTICE TASKS

Questions provide opportunities for developing skills that you have learnt about in the demo tasks.

CHALLENGE TASKS

Challenge tasks will stretch and challenge you even further.

SUMMARY

There is a summary of key points at the end of each chapter.

END-OF-CHAPTER TASKS

Questions at the end of each chapter provide more demanding programming tasks, some of which may require use of knowledge from previous chapters. Answers to these questions can be found in the solutions chapter, on the digital part of this resource.

NOTE: As there are some differences in the way programming statements are structured between languages, you should always refer back to the syllabus pseudocode guide to see how algorithms will be presented in your exam.

CAMBRIDGE IGCSE™ & O LEVEL COMPUTER SCIENCE: PROGRAMMING BOOK

> How to use this series

The coursebook provides coverage of the full Cambridge IGCSE, IGCSE (9–1) and O Level Computer Science syllabuses (0478/0984/2210) for first examination from 2023. Each chapter explains facts and concepts and uses relevant real-world contexts to bring topics to life, including two case studies from Microsoft® Research. There is a skills focus feature containing worked examples and questions to develop learners' mathematical, computational thinking and programming skills, as well as a programming tasks feature to build learners' problem-solving skills. The programming tasks include 'getting started' skills development questions and 'challenge' tasks to ensure support is provided for every learner. Questions and exam-style questions in every chapter help learners to consolidate their understanding.

The digital teacher's resource contains detailed guidance for all topics of the syllabuses, including common misconceptions to elicit the areas where learners might need extra support, as well as an engaging bank of lesson ideas for each syllabus topic. Differentiation is emphasised with advice for identification of different learner needs and suggestions of appropriate interventions to support and stretch learners.

The digital teacher's resource also contains scaffolded worksheets for each chapter, as well as practice exam-style papers. Answers are freely accessible to teachers on the 'supporting resources' area of the Cambridge GO platform.

There are three programming books: one for each of the recommended languages in the syllabuses – Python, Microsoft Visual Basic and Java. Each of the books are made up of programming tasks that follow a scaffolded approach to skills development. This allows learners to gradually progress through 'demo', 'practice' and 'challenge' tasks to ensure that every learner is supported. There is also a chapter dedicated to programming scenario tasks to provide support for this area of the syllabuses. The digital part of each book contains a comprehensive solutions chapter, giving step-by-step answers to the tasks in the book.

Chapter 1
Visual Studio Community

IN THIS CHAPTER YOU WILL:

- understand the two programming applications used in this book
- understand how to code and save a basic program in console application
- understand how to obtain input data and provide output in console application
- understand how to code and save a basic program in Windows Forms application
- understand how to use the main design windows in Windows Forms application
- understand the format of the event-driven subroutines used in a Windows Forms application.

Introduction

The process of creating computer program code has changed considerably since the early days of computer science, when the first programs were written directly in binary using punch cards.

Today we use high-level programming languages. These languages use coding statements that are close to human language. The code statements are then translated to binary before being executed by the computer. The majority of high-level programming is completed using an Integrated Development Environment (IDE), which provides the programmer with tools that help create, debug and translate the programs. Visual Studio 2019 Community is an IDE that helps to generate programs written using the Visual Basic programming language.

> **KEY WORD**
>
> **Integrated Development Environment (IDE):** software that helps programmers to design, create and test program code.

1.1 Getting Visual Studio 2019 Community for Windows

Visual Studio 2019 Community is the current version of free developer tools provided by Microsoft. The IDE supports a wide range of programming languages including Visual Basic, Visual C++ and Visual C#. Visual Basic provides an interface that allows students to develop programming skills while at the same time producing satisfying systems.

Visual Basic also provides programmers with access to a large class library. Classes are templates that hold prewritten code that support functionality of objects. As your skills increase, you will be able to use this feature-rich development environment to produce and publish complex systems. System requirements and download options can be found on the Microsoft website by searching 'Visual Studio downloads'.

> **KEY WORDS**
>
> **Visual Studio 2019 Community:** a version of the Integrated Development Environment (IDE) produced by Microsoft used to create programs using the Visual Basic programming language.
>
> **interface:** the way in which a user inputs data into and receives information from a computer system.
>
> **classes:** templates that hold prewritten code which supports the functionality of objects. All of the GUI elements in Visual Basic are objects of a class. The GUI objects can be created and manipulated without the need to write code as the code required is already contained in the class.

1.2 The Integrated Development Environment

> **Note:** Window Forms application is optional content. It is not covered in the syllabuses.

Visual Studio 2019 Community IDE is the free version of a full commercial programming IDE. It offers many different languages and different formats. This is useful when developing for different environments such as web design, mobile applications or computers. This book only uses two of the applications, **console application** and **Windows Forms applications**. Both make use of very similar coding approaches to algorithms but differ in how a user interacts with them. The features are shown in Table 1.1.

Application	Features of the application
Console application VB Console App (.NET Framework)	Used for creating a program that uses a **command-line interface** on a computer running a Windows operating system. This provides a textual interface uncluttered by the need to support a **Graphical User Interface (GUI)**.
Windows Forms application VB Windows Forms App (.NET Framework)	Used for creating a forms-based GUI program on a computer running a Windows operating system. Users interact with the program by using a GUI, which contains the type of elements we see on web pages and modern applications. These elements include areas to input values, buttons to select options and dropdown menus. The GUI can be customised by the programmer. We have included additional tasks throughout which require you to use GUI programming. This is not required by the syllabuses, but for your own interests and programming skills development. We have flagged this each time we include one of these tasks.

Table 1.1: Features of console application and Windows Forms application

Because console application is the required programming format at this level, most of the code for the demonstration tasks use console application. But all the solutions to the practice tasks are shown in both applications.

KEY WORDS

console application: a command-line interface provided by Visual Studio 2019 Community used to create programs using the Visual Basic programming language. This interface most directly matches the syllabuses.

Windows Forms application: a GUI provided by Visual Studio 2019 Community used to create programs using the Visual Basic programming language. Windows Forms application is an event-driven approach to programming where subroutines are linked to GUI objects.

command-line interface: an interface that uses text on a single screen to input data into a system and output information from that system.

Graphical User Interface (GUI): an interface that includes graphical elements, such as windows, icons and buttons.

1.3 Getting started

When you first open the IDE, you will see the window in Figure 1.1.

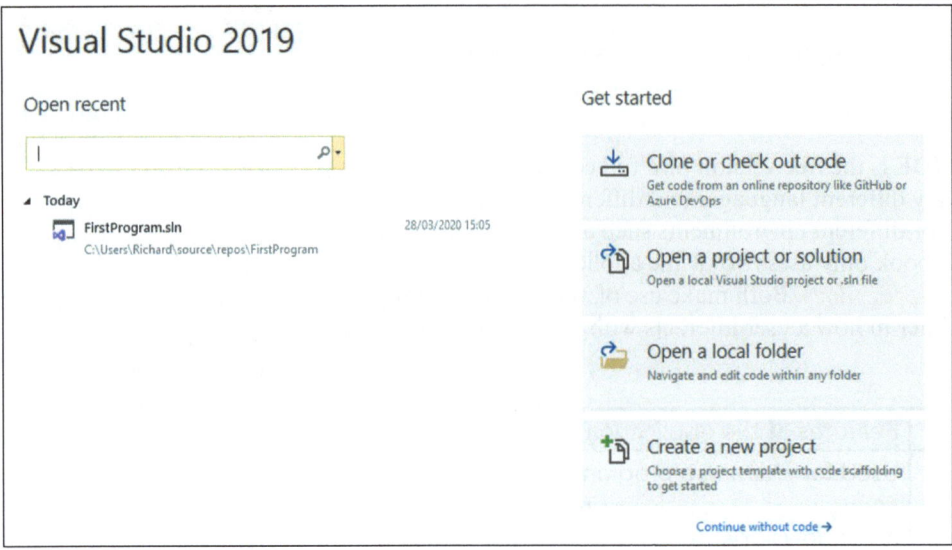

Figure 1.1: Visual Studio 2019 home page

The options you might use are described in Table 1.2.

Option	Description
Open recent	Allows direct opening of programs used recently. Alternatively, you can search for the recent programs used. NOTE: The file location shown in the example is the default repository folder 'source\repos', which is used for all Visual Studio 2019 programs. Your network administrator may have selected a different folder location.
Open a project or solution	Allows you to find and open a Visual Studio program stored in any folder or external storage device. This is useful if you are transferring programs between computers by using external storage devices.
Create a new project	Allows you to create a new program in console application or Windows Forms application.

Table 1.2: Options for using Visual Studio 2019

The 'Create a new project' window provides the option to create any of the many applications provided by the IDE. To reduce the available options, select Visual Basic from the languages drop-down menu, shown in Figure 1.2.

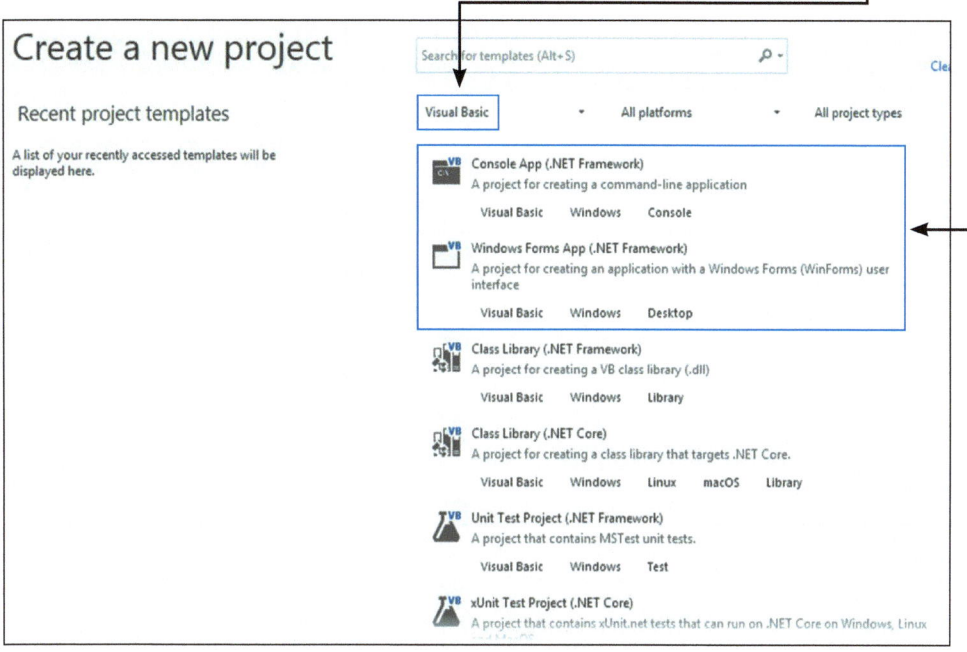

Figure 1.2: Create new project window

The two applications you will be using are shown in the list of available applications.

The applications may not appear at the top of the list as shown in Figure 1.2.

NOTE: Be careful when selecting the console applications. There is another console application for use with macOS and Linux operating systems on the list. This is called Console App (.NET Core)

You require Console App (.NET Framework)

1.4 Console application

When you start a new console application program, you will be prompted to name the project (see Figure 1.3). It is good practice to give all projects meaningful names.

Defining the name of the project using the 'Project name' box will change the Solution name to match the project name. It is possible to have several projects within one solution, but this is beyond the scope of this book.

This window allows you to select an alternative storage location for your project if you do not wish to use the default 'source\repos' folder.

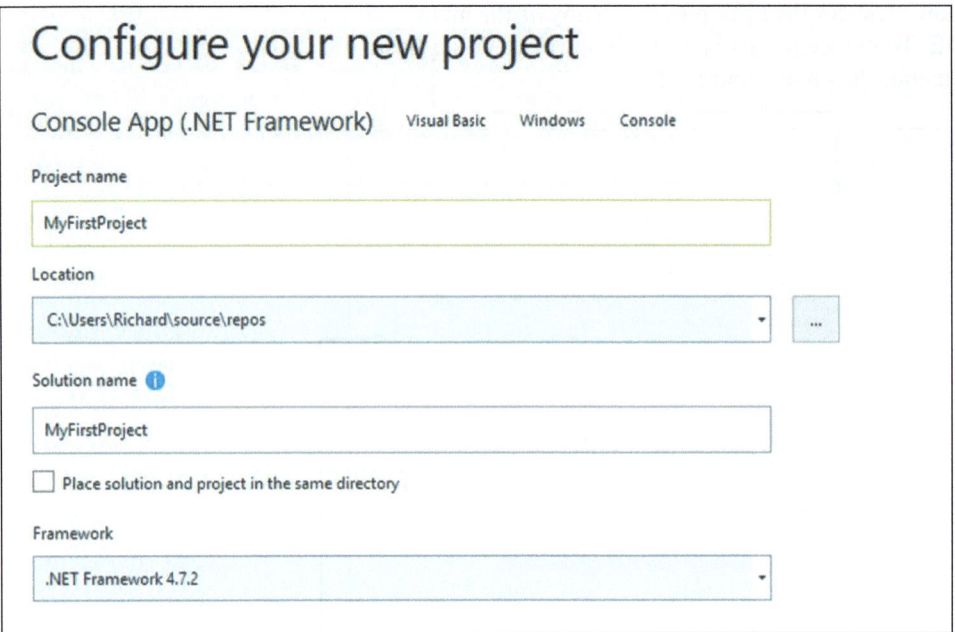

Figure 1.3: Configure your project window

The default layout of the console application consists of the main programming window (see Figure 1.4). This provides an area in which you write all the program code required to accept inputs, process data and display the required outputs. The Solution Explorer on the top right-hand side of the window displays the contents of a solution.

A solution holds all the components of a project. Our projects will be limited to a single module of code, which will be called Module1. You can see that shown in the Solution Explorer as Module1.vb. If you accidently close your code window you can reopen it by double clicking on Module1.vb in the Solution Explorer.

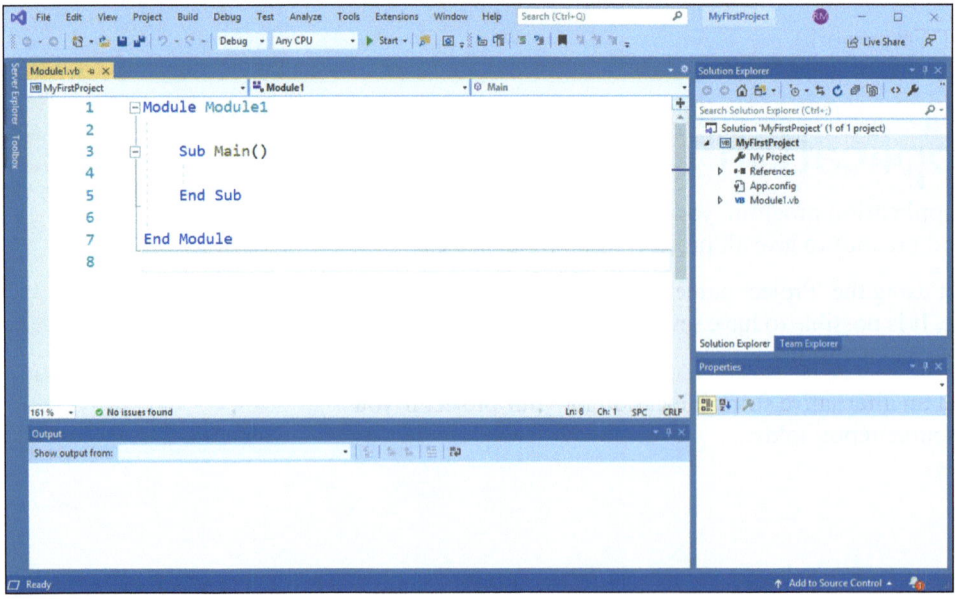

Figure 1.4: Console application programming window

Although it is much more complex than we will need, it is possible for Solutions to contain many modules and linked files. Solution Explorer provides a complete view of the files in a project and can be used to add or remove files and to organise files into subfolders. Remember, Visual Studio can be used for large commercial projects and Solution Explorer is very helpful in those types of project.

1.5 Make your first program using console application

When the console application is first loaded the code window will contain four lines of code.

```
Module Module1
    Sub Main()
    ...
    End Sub
End Module
```

Modules contain code and can hold a number of **subroutines** that perform specific actions. You will learn more about subroutines in Chapter 7. `Sub Main()` is the entry point for the program, and `End Sub` indicates the end of the subroutine. Code written between these points will be **executed** when the application is run. You can use the Enter key to add additional lines.

> **TIP**
>
> Do not change the name of the module `Module1` or the subroutine `Main( )`. These are the module and subroutine that the project will run when it is executed. If the module or subroutine does not exist, the project will not run.

Our first program will display the text 'Hello World'. You need to code the application to display the required text. This is the code we will be using. The explanations are in the steps below.

```
Module Module1
    Sub Main()
        Console.WriteLine("Hello World")
        Console.ReadKey()
    End Sub
End Module
```

KEY WORDS

subroutine: subroutines provide an independent section of code that can be called from another routine while the program is running. In this way, subroutines can be used to perform common tasks within a program.

execute: in Computer Science, the term 'execute' means the operation of a computer program. When a computer program is in operation it is being executed. The term 'run' is also used to describe the same process. The 'program is running' or the 'program is being executed' both mean the same thing.

Step 1a

```
Console.WriteLine("Hello World")
```

Type the word 'Console' into the Main subroutine shown in the code window (see Figure 1.5).

As you type you will notice that the IDE provides an autocompletion window listing all the code inputs or objects that match the letters you have typed. To autocomplete the entry, you can double click the correct item, or press the spacebar when the item is highlighted. This will speed up your coding as once you have typed in the first few characters of the instruction, the software will automatically highlight the closest match.

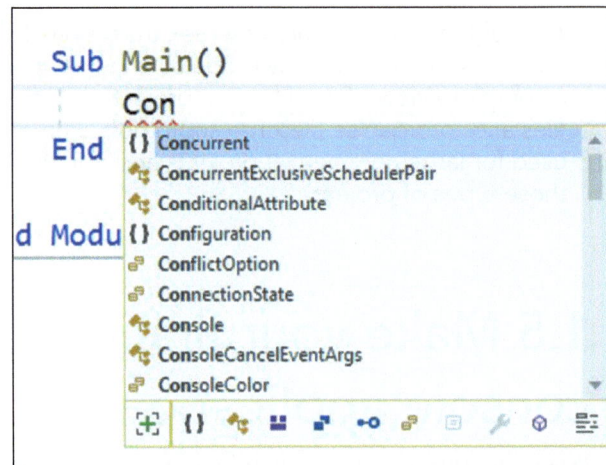

Figure 1.5: Autocompletion window

Step 1b

```
.WriteLine("Hello World")
```

In Visual Basic, the functionality of a class is accessed by use of the dot symbol. When the program reads inputs and displays outputs, it makes use of the **Console class**. The Console class provides access to a library of methods that allow the user to interact with the console.

When Console has been completed type a dot symbol. This will show a list of all the available methods for the Console class (see Figure 1.6). The method we need is 'WriteLine', which will display text in the console window when the code is executed. The method must be passed with the required text: `"Hello World"`.

NOTE: The text to be included is in quotation marks to indicate that it is text and not a reference to another object.

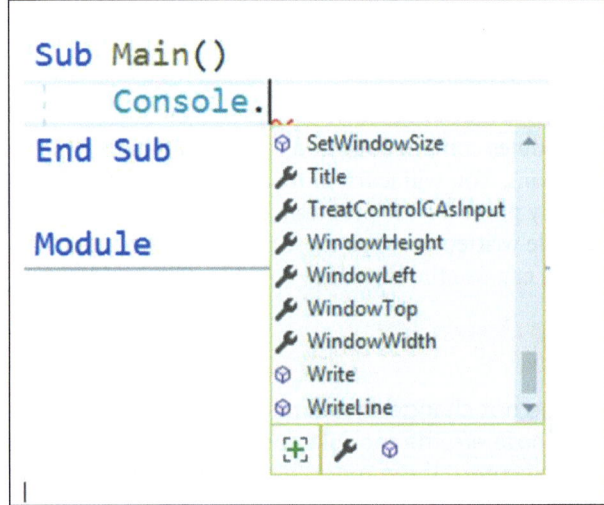

Figure 1.6: Methods list

Step 2

```
Console.ReadKey()
```

The code `Console.WriteLine("Hello World")` will display the required text in a console window when the application is run but the window will close as soon as the code has been executed. To prevent this, use the console 'ReadKey' method to pause the execution until a key is pressed on the keyboard.

1 Visual Studio Community

> **TIP**
>
> Typing `Imports System.Console` at the start of the program means you won't have to type console when you use the class:
>
> ```
> Imports System.Console
> Module Module1
>     Sub Main()
>         WriteLine("Hello World")
>         ReadKey()
>     End Sub
>
> End Module
> ```
>
> In this code, the Console class is imported at the beginning of the program. This means that rather than having to write `Console. ...` for both lines of code in `Sub Main()`, you can just write the methods needed (in this case, `WriteLine("Hello World")` and `ReadKey()`).

Step 3

To run the code, click the Start option on the toolbar ▶ Start ▾ or use F5 from the keyboard. This will launch the console window (Figure 1.7) and display the text 'Hello World'. The execution of the code will be halted until a key is pressed, at which time the window will close.

Figure 1.7: Console window

> While the console window is open, you may notice the Diagnostic Tools running in the background (see Figure 1.8). This provides information about the memory and CPU usage of the program being executed. While this is helpful when designing commercial projects, it is outside the scope of the syllabuses, and I usually shut the window.

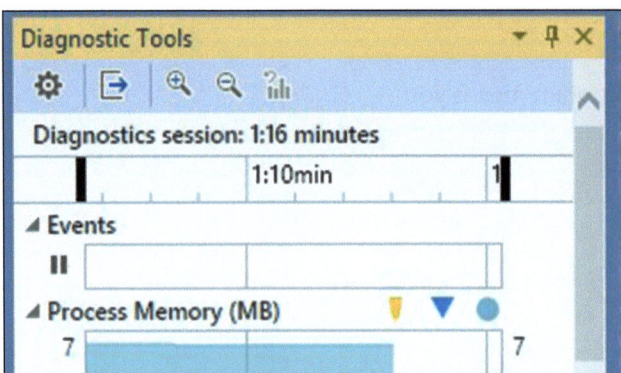

Figure 1.8: Diagnostic tools window

Congratulations! Your first console application program is complete.

Use the 💾 Save All option under the FILE menu to save your project.

> **Note:** Windows Forms application is optional content. It is not covered in the syllabuses.

1.6 Windows Forms application

Console application makes use of a single user interface to accept text-based inputs and display text-based outputs. Windows Forms application makes use of a range of **Graphical User Interface (GUI)** tools, to produce systems that have more in common with commercial applications.

Windows Forms application is more complex as programmers need to design and produce the graphical user interface that will allow the user to interact with the system. Visual Basic is an event-driven procedural language in which events trigger subroutines that execute the code within them.

In Visual Basic, user-controlled events – such as pressing the Enter key or selecting a button on a GUI – control the execution of the code. This is known as event-driven programming. A common element of a Windows Forms application GUI is a button that can be selected by a user. When the button is selected, it will trigger the execution of code held in a subroutine that is linked to the button. As you develop more complex projects your GUI will include many different buttons that a user can select, each button triggering a different section of code held in a subroutine linked to that button. For example, a reset button would be linked to a subroutine that holds the code required to reset the project.

In this first Windows Forms application, we will have a single button. Clicking this button on the form will trigger the code in the subroutine linked to the button that will deliver the message 'Hello World'.

To create a Windows Forms application, you will need to first create a new project (Figure 1.2) and select **Windows Forms App (.NET Framework)**. You should give your project a meaningful name and select an appropriate storage location when you configure your project (Figure 1.3).

The default layout of a Windows Forms application contains five main areas, these are shown in Figure 1.9 and described in Table 1.3.

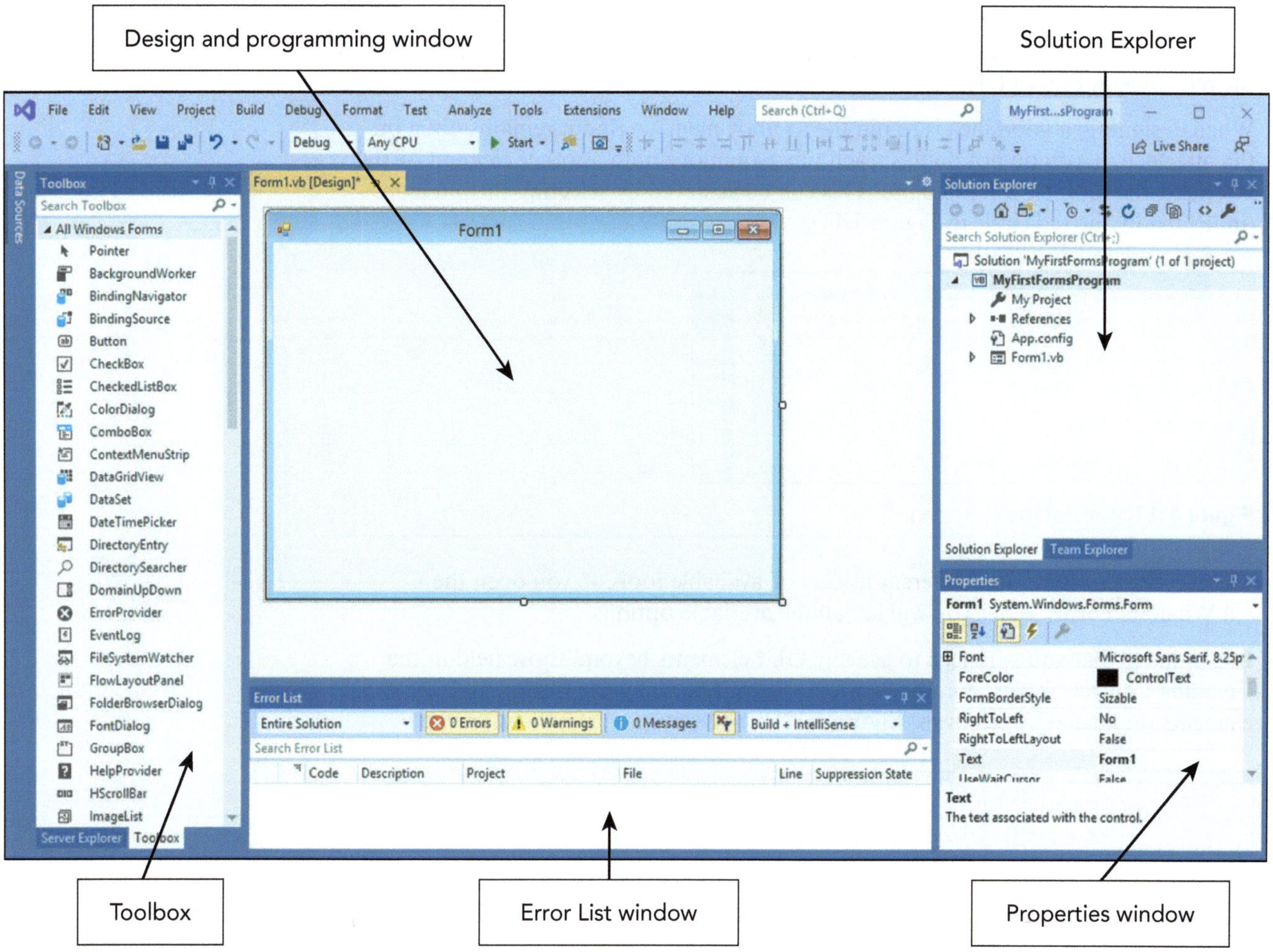

1 Visual Studio Community

Figure 1.9: Windows Forms application programming interface

If your Windows Forms application interface does not include all of these windows, or if you accidentally close a window, you can open additional windows by using the **View** option on the top ribbon menu.

An alternative to opening individual windows via the **View** option is to return to the default layout. This can be accessed via the **Window** option on the top ribbon menu (see Figure 1.10).

TIP

Individual windows on the interface can be sized, docked or set as floating. You can personalise the display of the interface to suit your own preferences. If you have dual-screen display, you can have different interface windows on different displays.

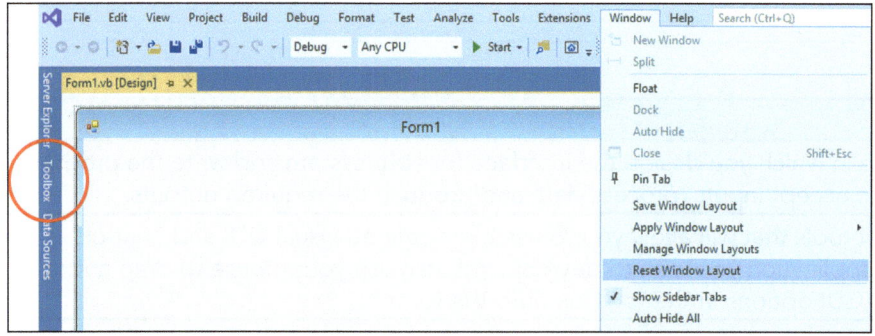

Figure 1.10: Default windows layout

11

It is likely that when you first access the interface the Toolbox will be minimised. If the toolbox is minimised, it can be seen on the left-hand side of the interface (indicated by the red circle in Figure 1.10).

When designing, a good idea is to pin the toolbox menu so that it remains open on the interface. If you do not do this, each time you move onto another element of the interface, the toolbox will minimise. This can be achieved by selecting the pin icon (indicated by the red circle in Figure 1.11).

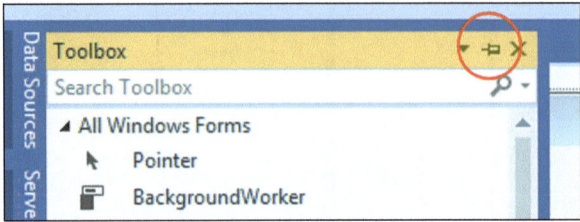

Figure 1.11: The Toolbox pin icon

The toolbox contains nine different folders of available tools. If you open the **All Windows Forms** folder you will see all the available options.

It is unlikely that you will need to use any GUI elements beyond those held in the **Common Controls** folder. For our first form we will be using the Button and Textbox elements (indicated by red ovals, Figure 1.12).

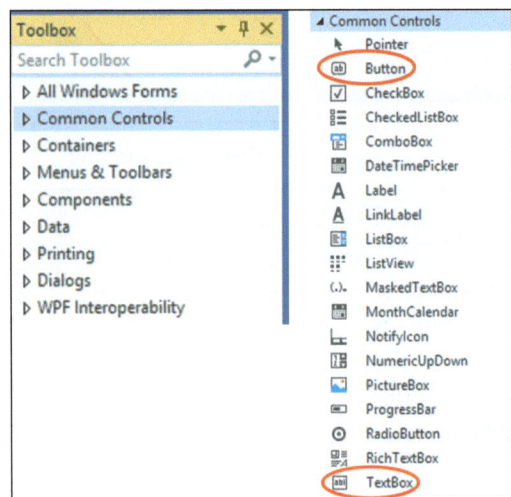

Figure 1.12: Toolbox folders

Window	Description
Design and programming	Provides an area in which you design the interface for your system and write the program code required to accept inputs, process data and produce the required outputs.
Toolbox	Provides a set of tools that will allow you to use a predefined visual GUI and control objects for the application you wish to develop. You can use your mouse to drag and drop any of the GUI options in the toolbox onto the form.

(continued)

Window	Description
Properties	Used to view and edit the configuration of selected objects. The Properties window shows the current properties of the current GUI object selected (for example, font type and size, colour, etc.). Changes to the properties of the object can be made by changing the options in the properties window. The properties of each element shown are those that are set before the program is executed. Any changes made during the execution of the program will be lost when the program is exited. For example, if a you wrote code to change the colour of a button when a user clicked on it, the button would return to the colour shown in the Properties window when the program was restarted. NOTE: The name of the GUI object currently selected is shown at the top of the window. Ensure you have selected the required object before making changes.
Solution Explorer	Displays the contents of a solution, which includes the forms and other project items. This is where you will find your forms and the program files that support them. It is possible to have multiple forms within one project; Solution Explorer will allow you to select individual forms.
Error List	Displays any errors, warnings or messages produced as you edit and execute code.

Table 1.3: Description of each window in the Windows Forms application programming interface

1.7 Make your first Windows Forms application

In a Windows Forms application, the 'Hello World' program will be achieved in two steps:

1. design and construct the user interface
2. code the program that will generate the required output.

NOTE: The code window is activated by double clicking on the form itself or any GUI object on the form. Soon we will deliberately activate the code window, but if you do this by accident, you can easily navigate back to the forms interface.

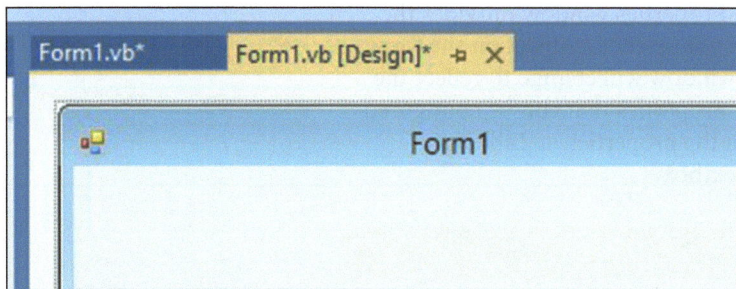

Figure 1.13: Code tabs

The code tab **Form1.vb** and design tab **Form1.vb (Design)** are both shown in the design and programming window (Figure 1.13), and selecting tabs allow you to swap between views.

Designing the interface

As with many other design applications (for example, website builders), the Windows Forms application allows you to design your graphical user interface using a drag and drop technique rather than writing code. You will have to code what the interface does later once you have designed it.

Find the button object [ab] Button in the Toolbox. Click to select the tool and move the mouse over the form in the main design window. The mouse icon will change to show the icon of the selected tool. Click and drag will generate a button on the form. Using the standard Windows mouse controls, it is possible to resize and move the button.

Use the same process to generate a textbox object [abl] TextBox on the form.

The form will initially look like Figure 1.14. We will customise this by changing the properties of the individual elements.

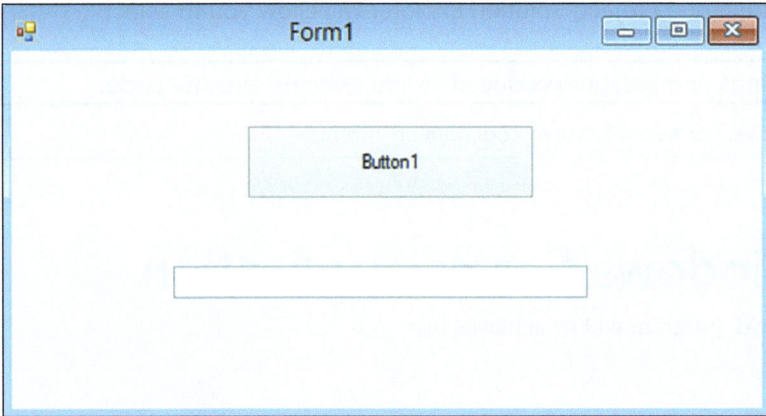

Figure 1.14: Initial form design

It is considered good practice to give objects meaningful identifiers. An identifier is the name of the object. The default identifier structure is the type of object followed by a number. The number increases depending on how many objects of that type are on the form, such as Button1, Button2, and so on. The Properties window provides the interface to change the properties of each object. As you select an object on the form, or the actual body of the form itself, the Properties window will change to reflect the properties of the object or form. Objects have many properties that can be configured by the designer, but we are initially only interested in the properties in Table 1.4. Use these properties to personalise the button and textbox.

> **TIP**
>
> When giving objects identifiers (names), it is good practice to start the identifier with capital letters that describe the type of object followed by a meaningful name. For example, BTNShow for the button used to show the 'Hello World' message. Putting the type description at the start will mean that when you type BTN on the code interface, the autocomplete will show all the buttons on your form. You could also start the identifiers of all textboxes with the letters TB followed by a meaningful name.

Property		Description
⊟ **Design**		The identifier (name) that will be used to refer to the object.
(Name)	**Button1**	
⊞ Font	Microsoft Sans Serif, 8.25p	The font style that will used to display text on the object.
ForeColor	ControlText	
Image	(none)	
ImageAlign	MiddleCenter	
ImageIndex	(none)	
ImageKey	(none)	
ImageList	(none)	The text (currently Button1) that will appear on the object.
RightToLeft	No	
Text	**Button1**	
⊟ **Appearance**		The background colour of the object.
BackColor	Control	

Table 1.4: Descriptions of properties being used

To change the properties of individual elements you must select the required element. You will know which element you have selected because the name will be shown at the top of the properties window (Figure 1.15).

Properties

Form1 System.Windows.Forms.Form

Figure 1.15: Properties identifier

NOTE: If you accidently double click on an element you will activate the code window. You can easily navigate back to the design window (see Figure 1.13).

Figure 1.16 shows the final design of the form.

The button name has been changed to BTNShow from the default value of Button1. The text on the button has been changed to 'Show Message' from the default value of 'Button1'. The size of the text has also been increased to 12 pt from the default value of 8.25.

The textbox name has been changed to TBMessage from the default value of TextBox1. The font size of the textbox has also been increased to 12 pt.

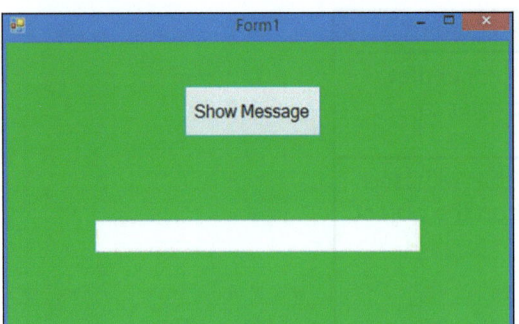

Figure 1.16: Customised form

The background colour of the form has been changed to green using the **Custom** colour pallet (Figure 1.17).

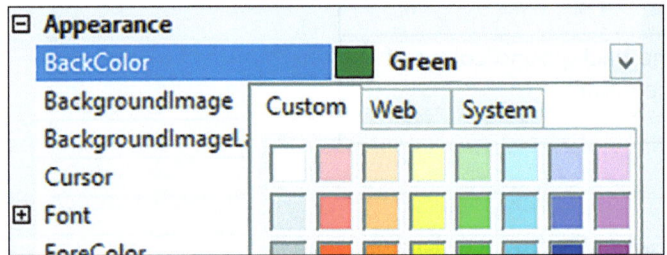

Figure 1.17: The BackColor property custom colour pallet

> **TIP**
>
> Make sure you have selected the correct element by checking the name at the top of the properties window. If you do not it is possible to change the properties of the wrong element.

Code the program

To create or edit the program code that will be activated by the form you have designed, you will need to open the code window.

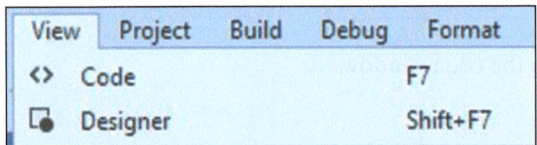

Figure 1.18: Opening the code window

To open the code window, select Code from the View menu (Figure 1.18). The code window will open as an additional tab in the main design window. Clicking the tabs will switch between design and code windows (Figure 1.19).

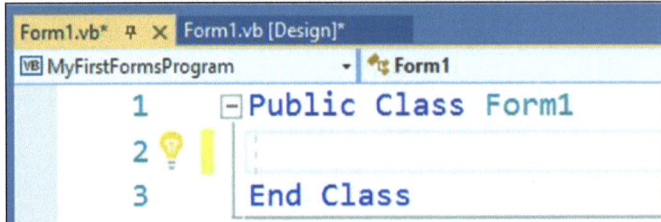

Figure 1.19: The code window

16

1 Visual Studio Community

The code window will contain only two lines of code. However, this code is important as it indicates the start and end of the code attached to the form. All additional code will be placed between these two indicators. Generally, code placed outside will not be part of the form and will cause an error. The Enter key can make additional lines.

In Visual Basic, user-controlled events – such as pressing the Enter key or selecting a button on a GUI – control the execution of the code. This is known as event driven programming. A common element of a Windows Forms application GUI is a button which can be selected by a user. When the button is selected it will trigger the execution of code held in a subroutine that is linked to the button.

In this first program, clicking the button on the form will trigger the execution of code held in a subroutine linked to the button that will deliver the message 'Hello World'. Before writing code, the event subroutine linked to the button has to be created.

The simplest way to create a subroutine is to double click the object on the interface you wish to use as a trigger (in this case, the button on the form). This will automatically create the code for the default subroutine attached to the object selected and show the code window.

```
Public Class Form1
    Private Sub BTNShow_Click(sender As Object,
    e As EventArgs) Handles BTNShow.Click
    End Sub
End Class
```

Let us examine the code to identify what each element achieves (Table 1.5).

> **TIP**
>
> Double clicking an object on the form to generate a subroutine can be done by mistake. This could mean that if you have accidentally double clicked on an object, your program will contain unused event subroutines. It is a common mistake to place the correct code in the wrong event subroutine. If your code does not run as expected, check you have used the correct event subroutine.

> **Note:** Although this makes use of object-oriented language and is outside of the scope of the syllabuses, it is useful to have some understanding of how the process works.

Code Element	Description
`Private Sub`	The start of an individual subroutine. `Private` means that the subroutine is only accessible by this form.
`BTNShow_Click`	The name of the subroutine. The automatic default is to name the routine after the object and event that will trigger the subroutine. However, it is possible to rename the subroutine.
`(sender As Object, e As EventArgs)`	The arguments, or data, that are associated with the event. As this is a button click event, the arguments are limited – either the button was clicked, or it was not. However, events associated with mouse activation, for example, will hold data about the location of the mouse on the form and which mouse button was clicked. You should not change or delete any of this data as your subroutine might not work. As you become a more advanced programmer, you will learn how you can manipulate these sections.
`Handles BTNShow.Click`	The event that will trigger the subroutine. In this case, clicking `BTNShow` will call the subroutine and execute the code it contains. It is possible to have a single subroutine triggered by multiple events.
`End Sub`	The end of the subroutine. All the code that is to be executed when the subroutine is called is placed between `Sub` and `End Sub`.

Table 1.5: Breakdown of code elements

We can now start to write some code. Within the Button Click subroutine, type in the identifier (name) of your textbox. In this example, the identifier of the textbox has been changed to TBMessage. As you type, you will notice that Visual Basic provides an autocompletion window (Figure 1.20) listing all the code inputs or objects that match the letters you have typed. You can double click the correct item or press the spacebar, to autocomplete the entry.

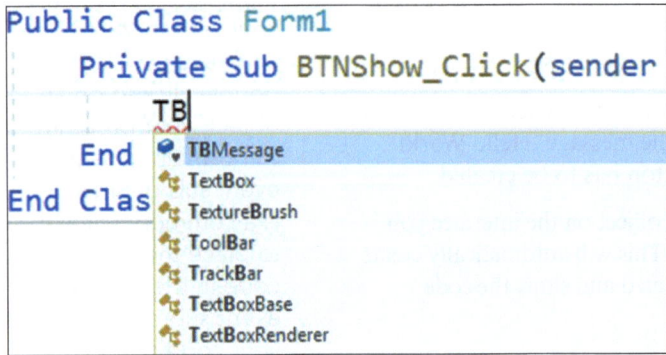

Figure 1.20: Autocomplete window

When the name of the textbox has been entered, type a dot symbol.

As the textbox is an object of the Textbox class, this will show a list of all the available methods that can be attached to a textbox object. The method we need is the Text method, which will either set text into a textbox or get text from a textbox. Fully type the word text, or alternatively double click or press the spacebar to select the method (Figure 1.21).

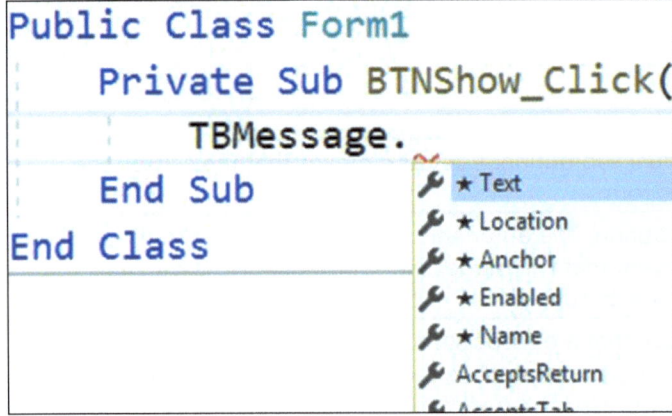

Figure 1.21: Autocomplete methods list

To indicate the actual text that should be displayed complete the code:

```
Public Class Form1
    Private Sub BTNShow_Click(sender As Object,
    e As EventArgs) Handles BTNShow.Click
        TBMessage.Text = "Hello World"
    End Sub
End Class
```

NOTE: The text to be included is in speech marks to indicate that it is new text and not a reference to another object.

1   Visual Studio Community

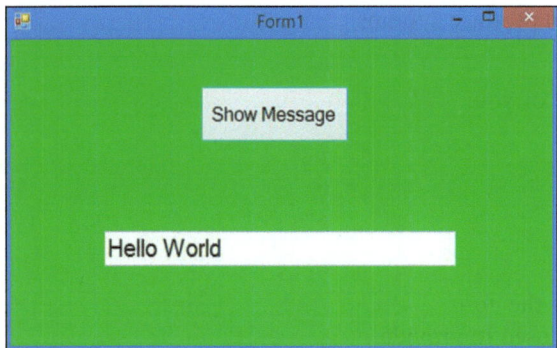

Figure 1.22: Hello World

To run the code, click the Start option on the toolbar ▶ Start ▼ or select F5 from the keyboard. This will launch the form as a separate interactive window. Click on the button and the text 'Hello World' will appear in the textbox (Figure 1.22).

In Figure 1.22, the background colour of the form has been changed. Also, the text on the button has been changed to read 'Show Message' and the font of both the button and textbox to be larger than the default 8.25 pt.

> Unlike a console application, the form will not automatically shut. It is a common mistake to try and alter code while the program is still running. To close a form, use the exit symbol in the top right corner of the form. Alternatively, select the Stop Debugging icon (Figure 1.23) on the main menu.
>
>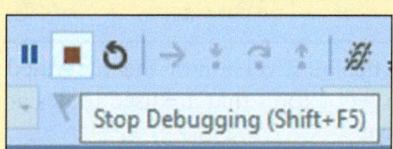
>
> **Figure 1.23:** Debugging icon

Your first Windows Forms application program is complete. You have made use of the design window and the Toolbox to create the interface. You have generated a subroutine called by the Click method of a Button object. Within that subroutine you have used the Text method of a Textbox object to place programmer-defined text into the textbox.

Use the 💾 Save All option under the FILE menu to save your project.

1.8 Choosing a console application or Windows Forms application

Throughout this book, the various tasks are completed using one or sometimes both of these applications. All the solutions are provided using both methods.

Console applications offer the benefit of reflecting more accurately the programming style of the syllabuses. Console applications do not involve the additional complexity of having to reference objects from GUI forms.

Windows Forms applications will offer a richer visual experience and produce systems similar to those commercially available.

Making use of both applications will best support the development of your computational thinking.

1.9 Additional support

This book aims to introduce programming concepts, making use of the non-language-specific formats included in the syllabuses. Visual Basic is used to provide the opportunity for you to use a real programming language to develop your understanding of these concepts. Additional support and guidance on the Visual Basic programming language and Visual Studio 2019 Community can be accessed directly from the Microsoft Virtual Academy.

A range of video tutorials and links to other support can be accessed from Microsoft's Visual Studio website. Just search for 'Microsoft Visual Studio getting started'.

SUMMARY

Visual Studio 2019 Community is the latest version of the integrated development environment IDE) produced by Microsoft. The IDE provides a set of tools that are used to design, create and debug programs using the Visual Basic programming language.
Visual Basic makes use of classes which hold prewritten code that supports the functionality of elements used. For example, the console class allows data to be read using `console.readline()` or displayed using `console.writeline`. All of the GUI elements in Visual Basic are objects of a class. The GUI objects can be created and manipulated without the need to write code as the code required is already contained in the class.
The Visual Studio 2019 Community IDE allows for programs to be created using either a console application or a Windows Forms application. • Console application is a text-based command-line interface, that most directly matches the syllabuses. The code is produced using a single programming interface. When the code is executed the user-interface is a single window used to input data into a system and output information from that system. User interaction with the program is text based. • Windows Forms application is a rich, event-driven programming environment that supports a graphical user interface (GUI). The design of a program requires the development of the GUI. Elements of the GUI are linked to subroutines in which code is written to perform the required tasks. When the program is executed the user interacts through the use of elements such as buttons and text buttons on the GUI. Information output from the system is also shown on the GUI. The GUI can be customised by the programmer to suit the needs of the system.
Subroutines are used in both applications to contain instructions relevant to a specific task. The subroutines can be triggered by user interaction such as pressing the Enter key or clicking a button on the GUI.

Chapter 2
Programming constructs

IN THIS CHAPTER YOU WILL:

- learn the difference between the three programming constructs: sequence, selection and iteration
- understand the role of flowcharts and pseudocode when designing programs
- understand the main symbols used in flowcharts
- learn about the preferred format of pseudocode.

Introduction

Many sporting activities make use of a limited range of approaches. Racket sports – such as badminton, for example – consist of a surprisingly limited number of different shots. Drop shots close to the net, long cleared shots to the back of the court and hard-driven shots account for a large proportion of the shots used in a game. The skill is in combining those shots effectively to win the game.

Programming also makes use of a limited range of **constructs** to achieve satisfactory solutions to complex tasks. Like a good badminton player, who combines different shots to win a game, an effective programmer will combine different constructs to produce efficient programs.

> **KEY WORD**
>
> **construct:** a method of controlling the order in which the statements in an algorithm are executed.

2.1 Programming concepts

Visual Basic and other procedural languages make use of three basic programming constructs (Table 2.1). Combining these constructs provides the ability to create code that will follow a logical process.

Sequence	The order in which the code is executed.
Selection	Providing options that will allow the execution (running) to follow a different path through the program based on certain criteria.
Iteration	Repeating a sequence of steps during the execution of the code.

Table 2.1: Programming concepts

In Computer Science, the term 'execute' means the operation of a computer program. When a computer program is operating, it is being executed. The term 'run' is also used to describe the same process. The 'program is running' or the 'program is being executed' both mean the same thing.

> **KEY WORDS**
>
> **sequence:** code is executed in the order it is written.
>
> **selection:** code follows a different sequence based on what condition is chosen.
>
> **iteration:** code repeats a certain sequence a number of times depending on certain conditions.

Sequence

The order in which a process is completed is often crucial to the success of that process. Take the mathematical expression $A + B \times C + D$. The rules of precedence dictate that the multiply operation ($B \times C$) will be completed first. Had the programmer wanted $A + B$ and $C + D$ to be completed before multiplying the two resultant values then they would have had to be explicit about the required sequence: $(A + B) \times (C + D)$.

To calculate the time it takes to complete a journey, you need to know the distance to be travelled and the speed you will travel. The first logical step would therefore be to calculate the distance to be travelled as, without this data, you could not complete the rest of the task.

Most tasks will involve different steps, and those steps often have to be completed in some logical sequence. When designing computer programs to achieve tasks, it is crucial to consider the sequence in which the task needs to be completed.

> **TIP**
>
> The mathematical rule of operation precedence, commonly known as BODMAS, also applies to operators used in programming in the same way it applies to normal mathematical calculations.

For example, when a satnav calculates an estimated arrival at destination time, the process follows a series of logical steps:

Step 1   Obtain location of journey start point.
Step 2   Obtain location of journey end point.
Step 3   Calculate shortest route between the two points.
Step 4   Calculate the time it will take to complete the journey.
Step 5   Obtain the time the journey will start.
Step 6   Calculate the estimated arrival time by adding the journey time to the start time.

Obtaining the journey start and end points is crucial because without this information, the rest of the process could not be completed. However, the order in which the two locations is obtained is not that important. Had Step 1 and Step 2 been reversed, the calculation at Step 3 could still have taken place.

In programming, the sequence is indicated by the order in which the code is written, usually top to bottom. The program will execute the first line of code before moving to the second and subsequent lines. An example of a sequence error would be completing a process before all the appropriate user inputs had been obtained. For example, trying to calculate the area of a rectangle when you only had the length of one side.

Selection

Often your programs will perform different processes dependent on user input. Consider a system designed to provide access to the school network when a user inputs a username and password. The system would need to follow different paths depending on whether the user input was accurate or not.

One path would be followed if the username and password input matched records that would allow network access. If the username and password input did not match the records, the system would follow another path in which the network access would be denied and the user prompted to re-input the details.

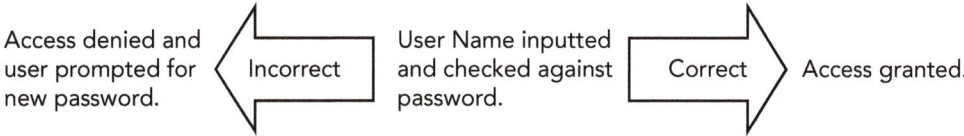

Figure 2.1: Different paths needed on input of username to access school network

Iteration

It is common for a program to perform identical processes on different data items. Consider a program that produces a line graph. It takes a series of coordinates and draws a line. The code that provides the instructions that plot the new coordinates and draws a connecting line from the previous coordinates will be repeated for each coordinate given.

Iteration allows the execution of the code to jump back to the beginning of the code sequence that provides instruction on how to 'plot' every time a new set of coordinates is entered.

2.2 Design tools

When you design programs, it is normal to plan the logic of the program before you start to code the solution. This is an important step in the design of effective systems because a flaw in the logic will often result in programs that run but produce unexpected outputs.

The first step in the design process is to break down the problem into smaller problems. This is called **top-down design**. Once you have the smaller problems defined then you can consider each problem separately. This will be easier to plan and finally code. You can use a structure diagram to help organise the top-down design. Chapter 6 explains top-down design and structure diagrams.

The next stage is to design an **algorithm** for the individual problems. Two approaches (shown in Table 2.2) that can be used at this stage to help create logically accurate systems are flowcharts and pseudocode.

Flowcharts	A graphical representation of the intended logic and sequence of a program or 'flow of a program' (which is where the name 'flowchart' came from). It is a diagram that describes the flow that the execution of code will take through a program.
Pseudocode	A method of describing the logic and sequence of a program using natural language to explain each step in the intended process. The logic can be understood by programmers without the need for detailed knowledge of any specific programming language. The term 'pseudo' is derived from the early Greek word for 'false'.

Table 2.2: Algorithm design tools

> **KEY WORDS**
>
> **top-down design:** a way of designing a computer program by breaking down the problem into smaller problems (subsystems) until it is sufficiently defined to allow it to be understood and programmed.
>
> **algorithm:** a process or set of rules to be followed during the execution of a program.

For this course, you will be expected to have a working understanding of both flowcharts and pseudocode and to use them to answer questions that ask you to explain the logic of your solutions to given tasks. Both methods are used throughout this book to indicate the logic of systems, and it is important that you become familiar with their use.

2.3 Flowcharts

Flowcharts are graphical representations of the logic of the intended system. They make use of symbols to represent operations or processes which are joined by lines that indicate the sequence of operations. Table 2.3 details the symbols used.

> **KEY WORD**
>
> **flowchart:** a graphical representation of the sequence and logic of a program.

Symbol	Notes	Examples
Terminator	The START or STOP of a system.	START STOP
Input or output	A required INPUT from the system user or an OUTPUT to the system user. The value being input or output is written on the symbol.	INPUT number OUTPUT result

(continued)

2 Programming constructs

Symbol	Notes	Examples
Process ▭	A process within the system. The flowchart should show sufficient detail to indicate how the proposed process is to be achieved. Beware of making the process too generic. For example, if the system had to calculate an average value, a process entitled 'Calculate Average' would be too generic. It needs to indicate the inputs or other values used to calculate the average.	result ← A * B average ← (A+B+C+D)/4
Logic flow ↓	Joins two operations. The arrowhead indicates the direction of the flow.	INPUT A, B → result ←A * B → OUTPUT result
Decision ◇	A point in the sequence where alternative paths can be taken. The **condition** on which the flow is determined is written within the symbol. Where multiple alternatives exist, sequence flows are indicated by chained decision symbols. Each 'FALSE' condition directs to another decision.	TRUE ← Number > 10 → FALSE → Is Input = A (FALSE/TRUE) → Is Input = B (FALSE/TRUE) → Is Input = C (FALSE/TRUE)

Table 2.3: Symbols used in a flowchart

DEMO TASK 2.1

Use a flowchart to represent the process of calculating an estimated arrival time for an aeroplane journey.

Solution

The steps in the process are:

Step 1 Obtain location of journey start and end airports.

Step 2 Obtain the distance between the two airports.

Step 3 Calculate the time it will take to complete the journey.

Step 4 Obtain the time the journey will start.

Step 5 Calculate the estimated arrival time by adding the journey time to the start time.

Step 6 Output the estimated arrival time.

KEY WORD

condition: the criteria that are tested as part of the execution of the code. The condition will result in either a True or False answer when tested.

> **CONTINUED**

```
                    ┌─────────┐
                    │  START  │
                    └────┬────┘
                         ▼
                 ╱───────────────╲
                ╱ INPUT Start Airport╲
                ╲ INPUT End Airport  ╱
                 ╲───────────────╱
                         ▼
                ┌─────────────────┐
                │ Obtain Journey  │
                │    Distance     │
                └────────┬────────┘
                         ▼
                ┌─────────────────┐
                │   Calculate     │
                │  Journey Time   │
                └────────┬────────┘
                         ▼
                 ╱───────────────╲
                ╱ INPUT Start Time ╲
                 ╲───────────────╱
                         ▼
                ┌─────────────────────┐
                │   Arrival Time ←    │
                │Start Time + Journey Time│
                └──────────┬──────────┘
                           ▼
                 ╱───────────────╲
                ╱ OUTPUT Arrival Time╲
                 ╲───────────────╱
                         ▼
                    ┌─────────┐
                    │  STOP   │
                    └─────────┘
```

Flowchart 2.1: Flowchart for calculating arrival time for an aeroplane journey

This task follows a simple linear process, so the flowchart that represents the process also follows a single path.

NOTE: Observe the use of the different symbols to indicate the inputs / outputs from the processes that need to be completed. When you come to code programs these will help to identify the need to obtain user input.

Flowchart design

A flowchart is often created in two stages.

Stage 1

Plan the logic and order of the system. The flowchart will show the intended sequence. It will include the inputs, outputs, decisions and processes required. At this stage, the process rectangle and decision diamonds are likely to include a text description of the processes or decision that must be made. There may be little, if any, detail provided about how the decisions or processes will be completed. This is known as high-level design and allows the programmer to check the logic of the system.

Stage 2

The decisions and processes will be added to provide details about how they will be completed. As the design develops, you will be able to consider the factors involved in each process and include a greater level of detail about each process. At this stage, a programmer may discover that they need additional inputs to be able to complete the process effectively. Wherever possible the final flowchart should include well-defined processes. Effective flowcharts avoid ambiguity.

The arrival time flowchart (Flowchart 2.1) includes examples of both Stage 1 high-level process and Stage 2 well-defined processes. Details of these processes are shown in Table 2.4.

Stage	Process
Stage 1 High-level process	Calculate time to complete journey While this description accurately describes the process, it does not define how to complete the process. It does not define what speed should be used to calculate journey time. Different aircraft fly at different speeds, so it is likely that the user would be expected to input the speed to be used. However, this step is not shown.
Stage 2 Well-defined process	Arrival Time ← Start Time + Journey Time This is a well-defined process. There is little ambiguity about how to complete the process. The ← symbol means 'becomes the result of'. In this example, the arrival time will be the result of adding the journey time to the start time of the flight.

Table 2.4: Examples of the two stages of flowchart creation

The best practice in symbol detail is shown in Table 2.5.

Symbol	Best practice suggestions
Input and output	The word INPUT or OUTPUT should be included in the symbol. As the symbol for inputs and outputs is the same, this helps to avoid confusion.
	The name of the data to be input or output should be descriptive. Avoid single letter descriptions, such as INPUT A.
Process	A description of the process must be included in the process symbol.
	Complex processes should be broken down into a series of simple steps, each step being represented by a different processes symbol.
	A process can have only a single outward flow. If a process had two outward flow lines, which one would be followed?
Decision	A decision will have two outward flows. Where more than one option exists, a sequence of decisions should be used.
	The condition that will control the flow of execution must be included in the decision symbol.
	It is crucial to label the outgoing flows TRUE and FALSE. If these are missing the direction of flow is not known.

Table 2.5: Best practice in symbol detail

PRACTICE TASK 2.1

How would the flowchart in Demo Task 2.1 need to change to show the calculation of journey time based on a user input of preferred speed?

Can you identify any other ambiguities in this flowchart?

DEMO TASK 2.2

Use a flowchart to represent a system that will output the results of an entrance examination for a university. The system requires the input of the maximum possible score in the examination and the score achieved by the candidate. Candidates who achieve 80% or more of the maximum score will be invited for an interview to decide if they will be offered a place at the university. Candidates who score less than 80% will be rejected.

CONTINUED

Solution

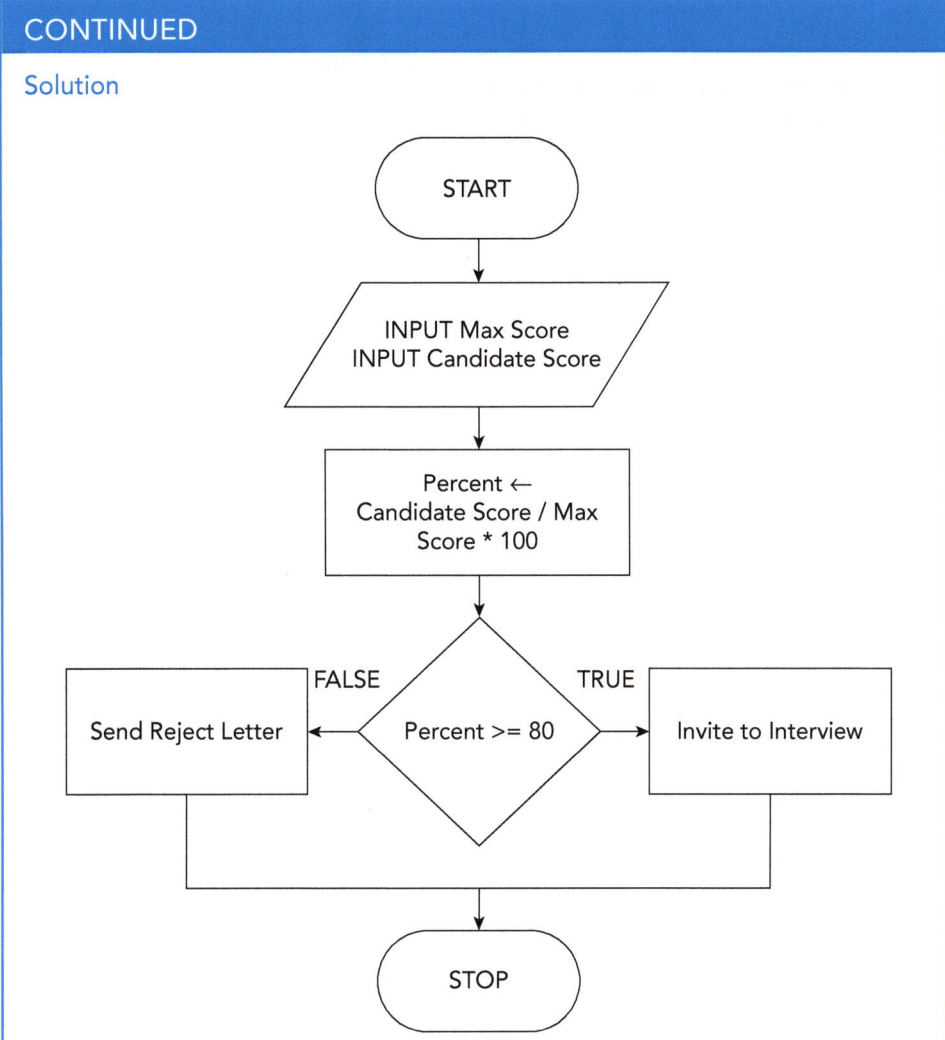

Flowchart 2.2: Flowchart to output the results of an entrance examination

NOTE: Observe the use of the diamond symbol to indicate a decision. The criteria are Boolean expressions that will either be TRUE or FALSE. It is crucial that the TRUE/ FALSE indicators are added to show how the execution of the code will flow.

PRACTICE TASK 2.2

How would Flowchart 2.2 need to change if the university introduced a process where candidates who achieved more than 90% in the entrance examination are automatically offered a place at the university without the need of an interview?

Redraw the flowchart to include this option.

DEMO TASK 2.3

Use a flowchart to represent a system that checks the entry of a new password. The user must enter the password twice. If the two entries match the system will accept the new password and output a message to the user. If the two entries do not match the system will reject the password and repeatedly prompt for it to be re-entered until the two password inputs match.

Solution

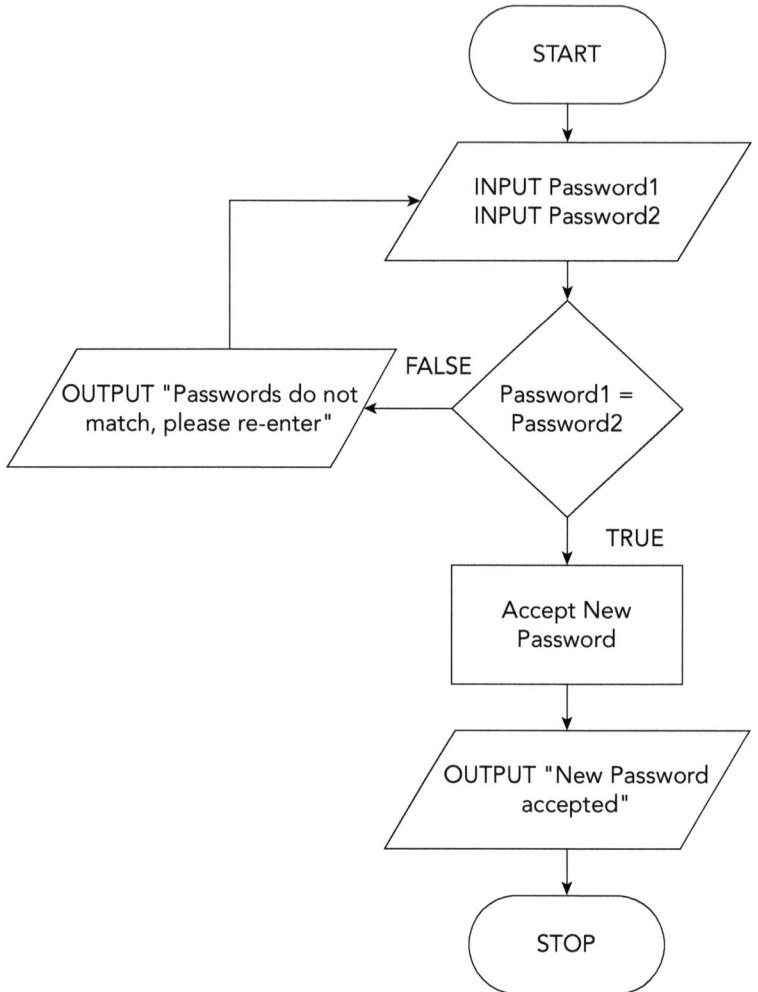

Flowchart 2.3: Flowchart to check the entry of a new password

NOTE: Observe the return flow from the FALSE path of the decision symbol. A flow that returns to an earlier stage in the flowchart represents iteration.

In this example, the return loop indicates the continual repetition of the steps:

- input password twice
- check passwords match
- if no match prompt for re-entry of passwords.

PRACTICE TASK 2.3

How would Flowchart 2.3 need to be changed if the requirements for the acceptance of the new password were altered to include the following?

- The password must contain more than six characters.
- The password must contain at least one symbol.

Redraw the flowchart to include this option.

You only have to include (Stage 1) general descriptions for this flowchart. You do not yet know how to program the checks required, so a general description of each check will be acceptable.

SKILLS FOCUS 2.1

COMMON FLOWCHART ERRORS

A system is required that will accept two numbers. The numbers must be larger than 10 and less than 20. Numbers outside of this range will be rejected and the user asked to re-input values. Once both numbers are within the range the system will output the result of adding the two numbers together.

Consider the attempt at creating a flowchart for the task shown in Flowchart 2.4.

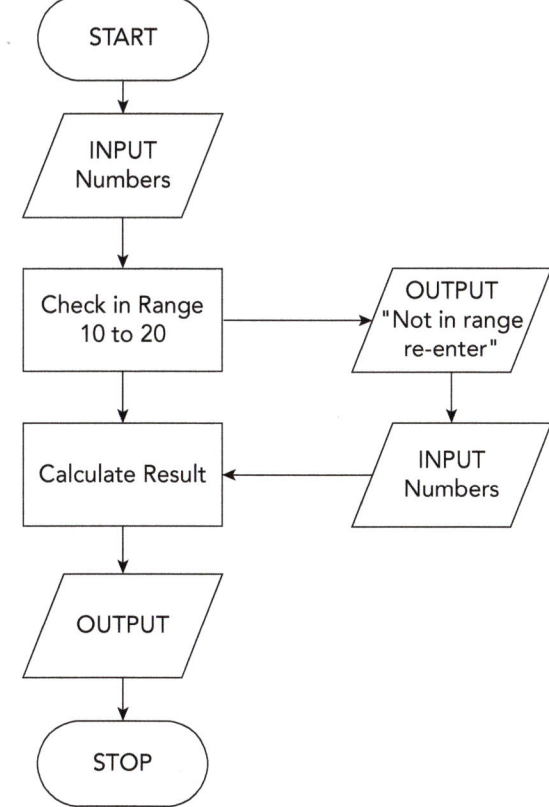

Flowchart 2.4: Flowchart attempt

> **CONTINUED**

The flowchart includes three errors as well as elements that could be better defined. Before you read on to the next part, which shows the errors and improvements, see if you can:

1 spot the errors and explain how to correct them

2 suggest the possible improvements.

Errors and possible improvements

While the flowchart may at first glance appear logical it does contain several errors and ambiguities. Table 2.6 identifies and explains the errors in the flowchart and Table 2.7 offers some improvements.

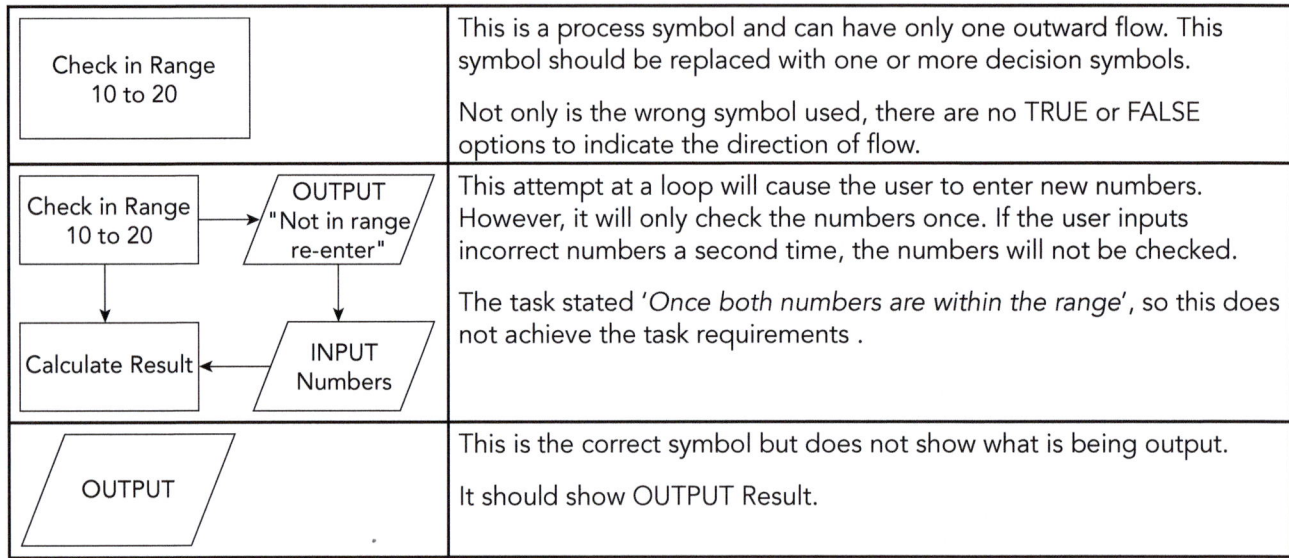

Check in Range 10 to 20	This is a process symbol and can have only one outward flow. This symbol should be replaced with one or more decision symbols. Not only is the wrong symbol used, there are no TRUE or FALSE options to indicate the direction of flow.
Check in Range 10 to 20 → OUTPUT "Not in range re-enter" → Calculate Result ← INPUT Numbers	This attempt at a loop will cause the user to enter new numbers. However, it will only check the numbers once. If the user inputs incorrect numbers a second time, the numbers will not be checked. The task stated 'Once both numbers are within the range', so this does not achieve the task requirements.
OUTPUT	This is the correct symbol but does not show what is being output. It should show OUTPUT Result.

Table 2.6: Errors in the flowchart

INPUT Numbers	This does not indicate how many numbers should be input. It would be improved by indicating the two numbers as individual inputs. Either by using two input symbols or by writing two INPUT descriptions in one input symbol.
Calculate Result	Although this process is accurately described, it is not well defined. It does not show what calculation should take place.
Check in Range 10 to 20	When replacing this incorrect symbol with decision symbols, the level of detail should be considered. Currently there is ambiguity in the words used. It does not define whether the check should include the values 10 and 20 or whether they should be excluded from the range. This can be made clearer through using the mathematical > and < symbols.

Table 2.7: Improvements

2 Programming constructs

> **CONTINUED**
>
> The corrected flowchart now looks like Flowchart 2.5.

> **TIP**
>
> When writing the detail in the decision symbols, make sure you include precisely defined mathematical conditions. These should be defined using appropriate mathematical symbols such as > or <. Using terms such as 'greater than' or 'less than' should be avoided.

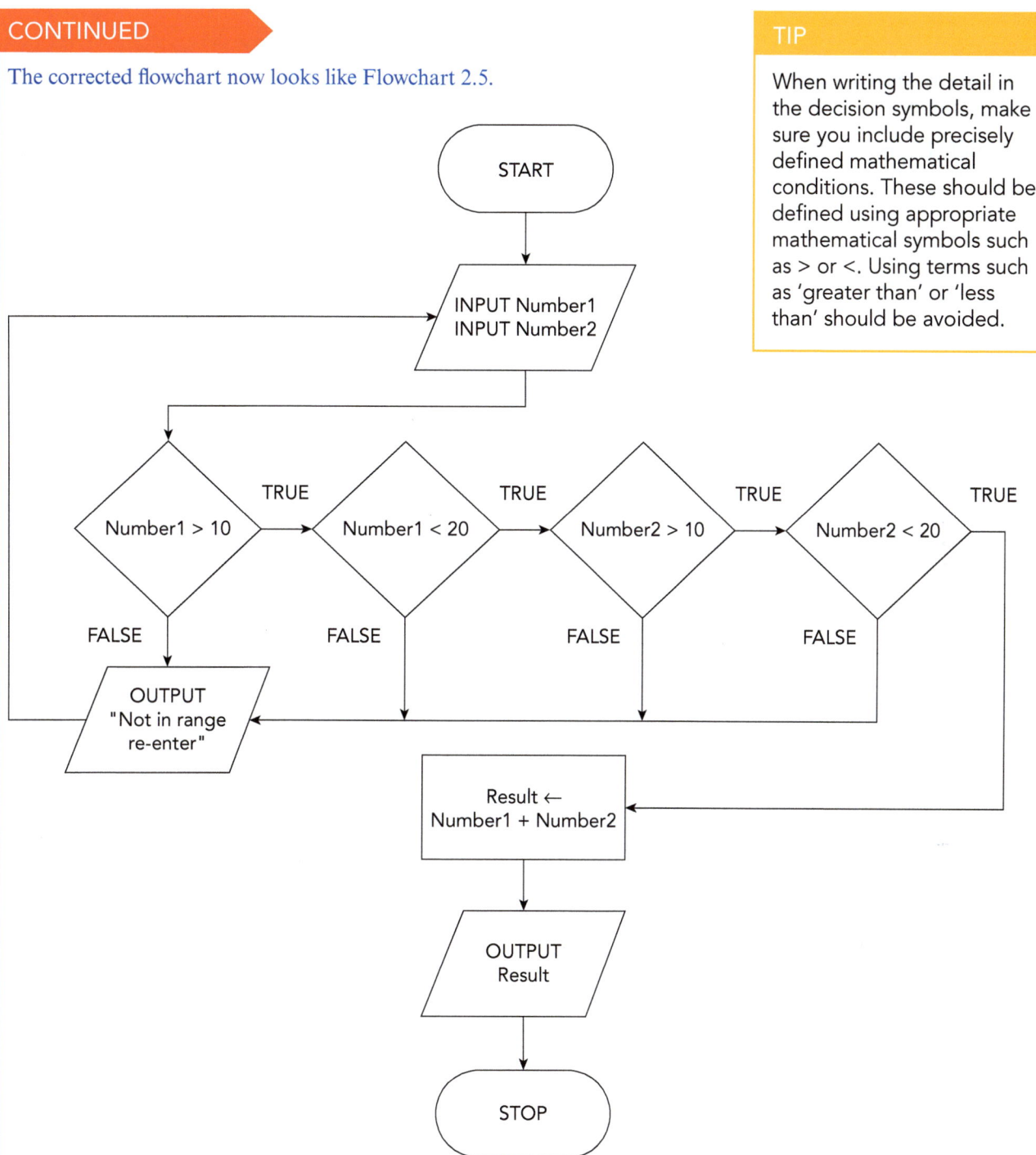

Flowchart 2.5: A correct flowchart for the two numbers task

The INPUT symbol now shows that two numbers need to be input.

The incorrect process 'Check in Range 10 to 20' has been replaced with a series of decision symbols. Each decision defines the range being checked precisely.

> **CONTINUED**
>
> Any incorrect entry is shown the error message and redirected back to the INPUT symbol. This process will continue to repeat until two in range values are input.
>
> The result process shows the exact process to be completed and the final OUTPUT symbol defines the value to be output.
>
> ### Questions
>
> A scenario question is likely to use words to describe acceptable range values. You will be expected to use precise mathematical terms to define the range in a flowchart.
>
> For the following scenarios, create an appropriate flowchart. Consider what inputs are required and what values need to be checked. Remember to be precise when defining the limits.
>
> 1 A roller-coaster has height and weight restrictions. To be able to ride the roller-coaster people must be at least 1.10 metres tall. The maximum weight a seat on the roller-coaster can safely hold is 125 kilograms. Only people who are within these limits will be permitted to rider the roller-coaster.
>
> 2 A firm makes metal beams. Beams come in two strengths 'standard' and 'extra strong'. The minimum length of beam they will make is 1 metre. A standard beam can be no longer than 4 metres, but an extra strong beam can be made in lengths up to 6.5 metres. Orders for beams outside of these limits will be rejected. All other orders proceed to a process called 'Calculate Price'.
>
> 3 A climate control system in a hospital operating theatre is required to maintain a constant temperature of 21° Celsius. A temperature sensor inputs the actual temperature in the operating theatre into the system. If the actual temperature drops below 20° Celsius, the heating must be on. If the actual temperature rises to 22° Celsius or more, the cooling system must be on. When the actual temperature is between these two limits both the heating and cooling system are off. Remember the temperature is constantly being checked.

2.4 Pseudocode

Pseudocode is a method of describing the logic and sequence of a system. It uses keywords and constructs similar to those used in programming languages but without the strict use of the syntax required by formal languages. It allows the logic of a system to be defined in a way that is independent of any specific programming language.

Pseudocode can be understood by any programmer. This means that the final program can be coded using any programming language that is appropriate to the context.

This is useful when designing programs as the code required to perform a particular process can vary considerably across differing programming languages. It also means that if there are multiple people working on a project, programmers from different programming language backgrounds can discuss and develop ideas and logic using a format common to them all.

> **KEY WORD**
>
> **pseudocode:** a way of unambiguously representing the sequence and logic of a program using both natural language and code-like statements.

Pseudocode follows a number of underlying principles:

- Use capital letters for keywords close to those used in programming languages.
- Use lowercase letters for natural language descriptions.
- Use indentation to show the start and end of code statements, primarily when using selection and iteration.
- Pseudocode should always include comments to explain the logic of the algorithm. A comment in pseudocode is defined by starting with the symbol //

One advantage of learning to program using Visual Basic is that the actual coding language is structured in a similar way to natural language and therefore closely resembles pseudocode. Visual Basic also automatically indents instructions where appropriate similar to the approach that should be adopted when writing pseudocode.

Pseudocode example

This pseudocode is for an algorithm that accepts the input of two numbers. These values are added together and the result is stored in a memory area called Answer. The value in Answer is then displayed to the user. (In Chapter 3, you will learn that this memory area is known as a variable.)

```
// Entering the values
INPUT Number1
INPUT Number2
// Calculate the addition and store in Answer
Answer ← Number1 + Number2
// Output the value in Answer
OUTPUT Answer
```

NOTE: Observe the use of ← to show the passing of values. This is distinct from the use of the equals symbol (=), which is used to indicate a comparison of two values. (Visual Basic does not have the ← symbol and uses the = symbol in both situations.)

2.5 Use of flowcharts and pseudocode in programming

Because of the universal nature of flowcharts and pseudocode, they are used extensively in the design of computer programs. The aim of this book is to help you to learn to design effective systems using the programming language Visual Basic. The following chapters make use of flowcharts and pseudocode to define the logic of systems, before moving on to specific Visual Basic-coded solutions.

Learning how to detail the logic of programs through the use of these design techniques will be a crucial step in your preparation for using the languages of the future. Language syntax is likely to change in the future but the need for effective logical and computational thinking will remain a constant.

CHALLENGE TASKS 2.1–2.2

2.1 A system is required by a teacher to calculate grades in an examination. The teacher will input the maximum marks that could have been achieve in the examination and the actual mark achieved by the student. The system will output the grade based on these boundaries:

    Grade as Distinction if percentage above 80

    Grade as Merit if percentage in range 61–80

    Grade as Pass if percentage in range 40–60

    Grade as Fail if percentage less than 40.

Design a flowchart to show the logic of the system required to output the correct grade.

2.2 A program is needed to work as a calculator. The user is asked to input two separate numbers and an arithmetic operation (+ , ×, ÷ or −). The program must then calculate the answer. This answer must be output.

Design a flowchart to show the logic of the system required to complete this task.

SUMMARY

Programmers make use of three constructs when writing code:
- sequence: the logical order in which the code is executed
- selection: branching of code onto different paths based on certain conditions
- iteration: repeating of sections of code.

Before coding a program it is crucial to design an appropriate algorithm using either a flowchart or pseudocode.

Flowcharts use a series of symbols to visually represent the logic of a program. The flow of the execution of the program is represented by flow lines with arrow heads to show the direction of flow. It is crucial that the correct symbols are used. The examination board will provide a flowchart guide and you must become familiar with the expected symbols.

Flowchart design can be done in two stages:
- the first stage is to consider the logic steps and processes required for the algorithm
- the next step is to consider how those steps could be achieved and ensure the flowchart contains enough detail to fully describe each step.

Decision symbols in flowcharts need to include precisely defined mathematical conditions. These should be defined using appropriate mathematical symbols such as > or <, for example, use A < B rather than A is less than B.

Pseudocode is a method of describing the logic of a program by using written language. The language used includes a series of key words that describe the exact programming constructs used in the program.

It is important that you comment your pseudocode (using the // symbol) to help explain the logic of your algorithm.

END-OF-CHAPTER TASKS

1. State the definitions of the following terms:
   - an algorithm
   - sequence
   - selection
   - iteration.

2. Explain what the following flowchart symbols are used for:

3. Create a flowchart to show the logic of the algorithm for this system:

   A user is required to input two numbers into the system. The second number must be exactly double the first number.

   The system will check that the second number is double the first. If this is not true, the system will output a message and the user must re-enter the numbers. This process will continue until the second number is exactly double the first.

   When this happens, the system will output the addition of the two numbers.

4. A business supplies sand to building sites. The delivery truck used has a maximum weight limit of 12 tons so no order greater than 12 tons can be accepted.

   Sand is provided in two different qualities:

Soft Sand	Charged $40.00 per ton (or part of a ton)
Sharp Sand	Charged at $36.50 per ton (or part of a ton)

   In addition to the cost of the sand a delivery charge of $20.00 is charged.

   The system will require the input of the amount and type of sand required. It will check the inputs meet limits and then output the total cost of the delivery.

   Create a flowchart to show the logic of the system required by the sand business.

Chapter 3
Variables and arithmetic operators

IN THIS CHAPTER YOU WILL:

- understand how to declare and use variables and constants
- be able to use the data types Integer, Real, Char, String and Boolean
- understand how to use basic mathematical operators to process input values
- understand how to design and represent simple programs using flowcharts and pseudocode
- understand the importance of commenting both pseudocode and programming code.

Introduction

In any subject, when working on complex tasks, you are likely to record the information you are using to help complete the task. For example, you might write down all the numbers used in a mathematical problem, updating the values as you work through each step in the problem. If you were completing a scientific experiment that involved recording the temperature of a chemical reaction every minute, you could write down a list of the temperatures and the time they were recorded.

Variables are the programming equivalent of your written records. They are small memory areas set up in your computer that can store data. The programs you write can use the variables to store values input by people using the program and to hold values that change as the program runs.

3.1 Variables and constants

Programs are normally designed to accept input data and process that data to produce the required output. Data used in programs can vary depending on the aim of the program; a calculator will process numerical data while a program designed to check email addresses will process textual data. When writing programs, you will use variables or constants to refer to these data values. A **variable** identifies data that can be changed during the execution of a program while a **constant** is used for data values that remain fixed. The **metadata** you provide about a variable or constant will be used by the computer to allocate a location in memory in which the data will be stored.

3.2 Types of data

To use a variable or constant, you must first give it an **identifier** (name). This is used as a label by the computer to reference the allocated memory. It is also important to provide information about the type of data so that the appropriate amount of memory can be reserved. For example, storing a large decimal number will require more memory bytes than storing a single character.

To support this process, different data types exist. The basic data types you will need to use are identified in Table 3.1.

> **KEY WORDS**
>
> **variable:** a memory location used to store a value; the value of the data can be changed during program execution.
>
> **constant:** a named memory location used to store a value; the value can be used but not changed during program execution.
>
> **metadata:** data about data; information about the structure or format of the data stored.
>
> **identifier:** the name given to each variable or constant. The identifier is used to reference the memory area.

Data type	Description and use	Visual Basic
Integer	Whole numbers, either positive or negative. Used with quantities such as the number of students at a school – you cannot have half a student.	Can store values ranging from – 2147483648 to 2147483647. Uses 4 bytes of memory. If a decimal value is put into an Integer variable, the value is rounded to the nearest whole number.

(continued)

Data type	Description and use	Visual Basic
Real	Positive or negative fractional values. Used with numerical values that require decimal parts, such as currency. Real is the data type used by many programming languages.	Visual Basic does not use the term Real. The equivalent data type is called 'Decimal'. The range of values depends on the number of decimal places required. Uses 16 bytes of memory. Stores a much larger range of numbers than the Integer data type. Single and Double also hold fractional numbers.
Char	A single character or symbol (for example, A, z, $, 6). A Char variable that holds a digit cannot be used in calculations.	Stores a single Unicode character. Uses 2 bytes of memory.
String	More than one character (a string of characters). Used to hold words, names or sentences.	Can store up to approximately 2 billion Unicode characters.
Textual data that is passed to a Char or String variable must be placed in straight double quotation marks (for example, "Hello", "a"). The quotation marks indicate that the value is textual data. If the quotation marks are omitted, the program will interpret the data as the identifier of another variable. For example, "Number_One" would be read as simple text, while Number_One without the quotation marks would be read as a reference to a variable named Number_One.		
Boolean	One of two values, either TRUE or FALSE. Used to indicate the result of a condition; for example, in a computer game a Boolean might be used to indicate if a player has achieved a higher level.	The default value for a Boolean variable in Visual Basic is FALSE. If you wish the variable to start with a TRUE value, the variable must be initialised before use.

Table 3.1: Basic data types

You can find more information about the data types in Visual Basic (many of these are outside the scope of the syllabuses) if you search for 'Data Type Summary Visual Basic' on the Microsoft website.

3.3 Pseudo numbers

Telephone numbers and ISBN numbers both consist of digits but are not truly numbers. They are only a collection of digits used to uniquely identify an item. Sometimes they contain spaces or start with a zero, and they are not intended to be used in calculations. These are known as 'pseudo numbers', and it is normal to store them in a String data type. If you store a phone number as an Integer, any leading zeros will be removed and no spaces or symbols will be permitted.

3.4 Declaring variables and constants

To **declare** a variable or constant, you will need to select an identifier (name) and a data type. In Visual Basic, it is possible to declare variables without declaring a data type, but it is considered good practice to define the data type. This is known as 'strong typing'; it allows the compiler to check for data type mismatches and results in faster execution of the code.

When naming your variables or constants, use identifiers that have a meaningful link with the data being stored. For example, if you are storing the high score of two players in a game, good names would be *Player1HighScore* and *Player2HighScore*. Declaring variables with relevant identifiers will help to make your code easier to read and maintain.

It is not possible in Visual Basic to have spaces in identifiers. It is common practice to start single word identifiers with a capital letter. If your identifiers use more than one word, it is normal to start each new word with a capital letter, as in this example, *Player1HighScore* and *Player2HighScore*. This is known as **PascalCase**.

Identifiers can include digits – for example, Number1 – but they cannot begin with a digit.

Terms that are used by Visual Basic for coding are known as 'reserved words' and cannot be used as variable names. For example, it is not possible to name a variable 'Integer' or 'Boolean'.

It is important to select the most appropriate data type for your variables. Using inappropriate data types may result in your programs returning unexpected results. For example, using the data type `Integer` to store currency values could result in the decimal element being rounded and the wrong values being output.

It is also considered good practice to give variables an initial value when declaring them: this is known as '**initialising**'. Uninitialised variables will hold the default values set by Visual Basic.

> **KEY WORDS**
>
> **declaring:** setting up a variable or constant.
>
> **PascalCase:** a way of creating a variable name from a combination of at least two words. Each new word starts with a capital letter.
>
> **initialising:** giving a variable a start (initial) value when it is first declared.

Declaring variables in Visual Basic

Declaration of variables is achieved by using the code format shown in Figure 3.1:

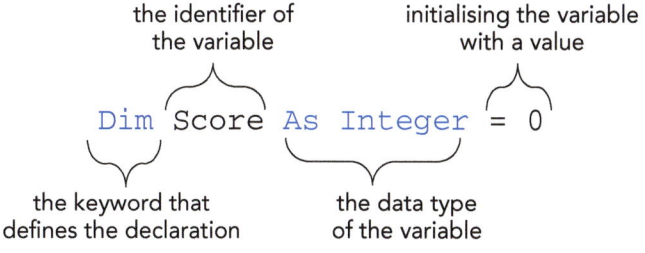

Figure 3.1: Declaring variables in Visual Basic

In the following example declarations, the variable identifier, data type and initial value are declared.

```
Dim PlayerName As String = ""
Dim DrivingLicence As Boolean = False
Dim Payment As Decimal = 0
```

Declaring constants in Visual Basic

The code for declaring constants follows a similar format to that used for variables, but uses the keyword 'Const' to replace 'Dim'. A constant must always be given an initial value.

Here is an example of declaring a constant:

```
Const Pi As Decimal = 3.14159
```

3.5 Variable scope

Where you place your variable or constant in the code will determine which elements of the program are able to make use of it.

Global variables can be accessed from any routine within the program. They are used for variables that need to be accessed from many elements of your program. To be a global variable or global constant, it must be declared outside of any specific subroutine. It is good practice to make all the global declarations at the start of the code.

Local variables can only be accessed in the code element in which they are declared. They are used when the use of the variables will be limited to a single routine. Using local variables reduces the possibility of accidentally changing variable values from other code elements.

> **KEY WORDS**
>
> **global variable:** a variable that can be accessed from any routine within the program.
>
> **local variable:** a variable that can only be accessed in the code element in which it is declared.

> **Note:** Windows Forms application is optional content. It is not covered in the syllabuses.

The following examples show the same code in a console application and in a Windows Forms application. There are two global variables (Score and PlayerName) and one local variable (Result).

Console application example

```
Module Module1

    Dim Score As Integer = 0
    Dim PlayerName As String = ""

    Sub Main()
        Dim Result As Integer = 0
    End Sub
End Module
```

Windows Forms application example

```
Public Class Form1

    Dim Score As Integer = 0
    Dim PlayerName As String = ""

    Private Sub Button1_Click(sender As Object, e As EventArgs) Handles Button1.Click
        Dim Result As Integer = 0
    End Sub
End Class
```

You can see that the process of declaring variables and constants is the same irrespective of the application mode used (e.g. `Dim Score As Integer = 0`).

3.6 Arithmetic operators

There are a number of operations that can be performed on numerical data. If you combine these operations and appropriate variables, you can create programs that are capable of performing numerical computation tasks.

The basic operators used in Visual Basic are shown in Table 3.2.

Operation	Example of use	Description
Addition	`Result = Number1 + Number2`	Adds the values held in the variables `Number1` and `Number2` and stores the result in the variable `Result`.
Subtraction	`Result = Number1 - Number2`	Subtracts the value held in variable `Number2` from the value in variable `Number1` and stores the result in the variable `Result`.
Multiplication	`Result = Number1 * Number2`	Multiplies the values held in variables `Number1` and `Number2` and stores the result in the variable `Result`.
Division	`Result = Number1 / Number2`	Divides the value in variable `Number1` by the value in `Number2` and stores the result in the variable `Result`.
Power of	`Result = Number1 ^ Number2`	Raises the value in variable `Number1` to the power of the value in `Number2` and stores the result in the variable `Result`. For example, if a user inputs 5 for `Number1` and 10 for `Number2`, the value passed to `Result` would be 5 to the power of 10, or 5^10.

Table 3.2: Basic operators used in Visual Basic

3.7 More complex arithmetic operators

Programming also makes use of integer division. The result of integer division is a whole number with a remainder. This can be useful when calculating ratios. Consider this example:

A school trip is being organised for 230 students. The students will be transported by bus. The school can book different size buses. The largest bus cost $400 and can carry 52 students. A smaller bus is also available that can carry 25 students and costs $250. What is the most cost effective way to transport all 230 students?

It is clear that if it is full the larger bus is the most economical but for 25 students or less the smaller bus is cheaper. The best option would be to hire 4 large buses to transport 208 students and a smaller bus to transport the remaining 22 students.

Using standard division 230 / 52 would produce 4.4231. This would show that at least four large buses where required but not output the amount of students still needing transport.

This is often thought of as the type of division you may have completed when you first started mathematics. For example:

230 integer division 52 = 4 remainder 22

Integer division involves two values:

- Quotient: an integer value (whole number) that is the number of divides possible – in our example, the quotient is 4.
- Modulus: an integer value that is the remainder following the integer division – in our example, the modulus is 22.

> **TIP**
>
> As a division operation can result in a fractional value, it is best to use a `Decimal` (or `Real`) data type to hold the `Result`.
>
> The mathematical rule of operation precedence commonly known as BODMAS also applies to operators used in programming in the same way it applies to normal mathematical calculations.

Table 3.3 shows some examples of the use of quotient and modulus.

Operation	Example of use	Description

Quotient	`Result = Number1 \ Number2`
	
	NOTE: Observe the use of the forward sloping slash, normal division uses a back-sloping slash (/)
	
	The pseudocode for a quotient division uses the term DIV
	
	The pseudocode for the programming statement above would be
	
	`Result ← Number1 DIV Number2`
		Divides the value in variable `Number1` by the value in `Number2` using integer division.
		
		Stores the exact number of divides in the variable `Result`.

Modulus	`Result = Number1 MOD Number2`
	
	The pseudocode for a modulus division also uses the term MOD
	
	The pseudocode for the programming statement above would be
	
	`Result ← Number1 MOD Number2`
		Divides the value in variable `Number1` by the value in `Number2` using integer division.
		
		Stores the remainder of the integer division in the variable `Result`.

Table 3.3: Explanation of quotient and modulus

In some programming languages the operators for MOD and DIV are library functions. In other programming languages they are just operators. For example, 2 MOD 4 is the same as MOD(2,4).

DEMO TASK 3.1

Using integer division, convert 500 minutes into hours and minutes.

Solution

Integer division can be particularly useful when converting between different representations of the same data. In this case, we would use integer division by 60. The value 60 is used as there are 60 minutes in one hour.

Hours would be obtained from the quotient outcome of the integer division. The remaining minutes would be obtained from the modulus outcome of the integer division.

Hours ← Minutes \ 60

Remaining Minutes ← Minutes MOD 60

Leaving us with the result:

500 \ 60 = 8 and 500 MOD 60 = 20, so 500 minutes is 8 hours and 20 minutes.

PRACTICE TASK 3.1

Think of other appropriate examples where you could use integer division to output a different representation of the same data. Identify the values that should be used to achieve the required output.

3.8 Programming

SKILLS FOCUS 3.1

MULTIPLY MACHINE

The Multiply Machine task will help you to develop programming skills by showing different methods of obtaining and storing user input in appropriate variables. It will also develop skills in using mathematical operators to calculate values before outputting them to the user.

The Multiply Machine takes two numbers input by the user, multiplies them together and outputs the result.

The program will need to complete a series of steps to achieve this task. You will need to consider the appropriate steps before you design the algorithm.

1. The program will need variables to hold the values input by the user.
2. The program will need to store the values input into the appropriate variable.
3. The program will need to complete the multiplication process and store the resultant value.
4. The program will have to display the resultant value to the user.

> **TIP**
>
> The order of the algorithm follows the same order as the steps. It is important to consider the sequence of the steps needed to complete a task when planning an algorithm.

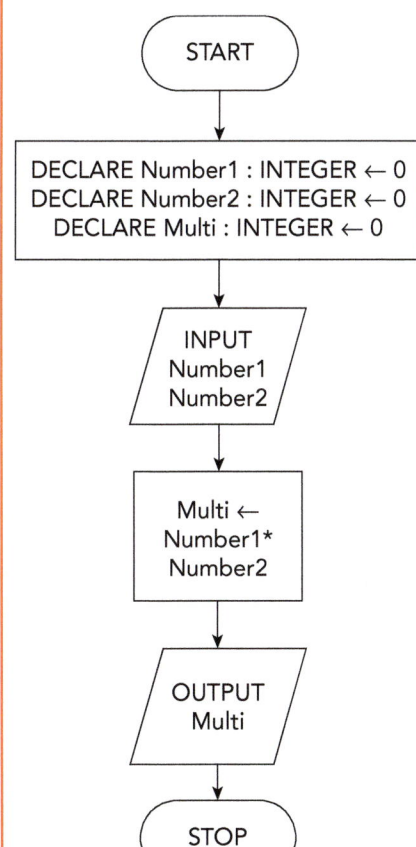

NOTE: Observe the use of the correct and recognised symbols in the flowchart and pseudocode:

- ▭ indicates a process such as completing the multiplication.
- ▱ indicates either an INPUT or OUTPUT.
- ← indicates the assigning of a value; it is used when initialising variables or passing new values to a variable.

Flowchart 3.1 shows the task and here is the pseudocode.

```
//Declaring the required variables
DECLARE Number1 : INTEGER ← 0
DECLARE Number2 : INTEGER ← 0
DECLARE Multi : INTEGER ← 0
// Obtaining the user input
INPUT Number1
INPUT Number2
//Calculate the multiple and store in variable
Multi ← Number1 * Number2
//Output the value of the multiplication
OUTPUT Multi
```

Flowchart 3.1: Flowchart for multiplication algorithm

> **CONTINUED**
>
> NOTE: Observe how both the flow chart and pseudocode represent the declaration of variables. It is important that the variables required are identified when designing a program. It is also important to state the initial value of the variable at the same time. In Visual Basic, variables are initialised with a default value if the initial value is not explicitly stated. For example, an integer variable is given a default value of zero. Therefore, it is particularly crucial to initialise variables with an initial value when the initial value required is not the default value.
>
> In Visual Basic, assigning is indicated by the use of the = symbol. In pseudocode, the ← symbol is used.
>
> > Variables that have been declared with numerical data types such as Integer or Decimal can only accept numerical data.
> >
> > If you input textual data for these variable types, the software will cause an exception error (an example is shown in Figure 3.2).
> >
> >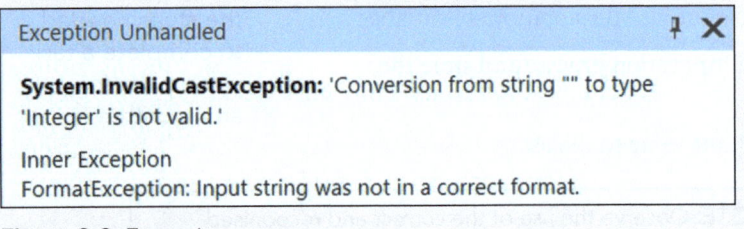
> >
> > **Figure 3.2:** Exception error
> >
> > A null input will also cause this error.
>
> Here is the console application program implementation of this solution:
>
> ```vb
> Module Module1
>
>     Sub Main()
>         'Declaration and initialising of required local variables
>         Dim Number1 As Integer = 0
>         Dim Number2 As Integer = 0
>         Dim Multi As Integer = 0
>
>         'Display a request for the first number
>         Console.WriteLine("Please insert first number")
>         'Store the user input into the Number1 variable
>         Number1 = Console.ReadLine()
>
>         'Request and store second user input
>         Console.WriteLine("Please insert second number")
>         Number2 = Console.ReadLine()
>
>         'Storing in the local variable the multiplication result
>         Multi = Number1 * Number2
> ```

CONTINUED

```
        Console.WriteLine("The answer is")
        'Displaying the value held in the variable Multi
        Console.WriteLine(Multi)


        'ReadKey used to pause the console window
        Console.ReadKey()
    End Sub
End Module
```

Note: Windows Forms application is optional content. It is not covered in the syllabuses.

If you are using the Windows Forms application, you will need to design an interface that is capable of taking two values and displaying a result. It could look something like Figure 3.3.

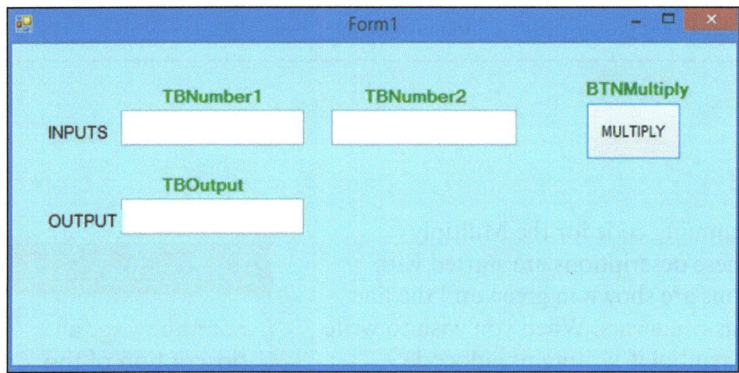

Figure 3.3: Windows Forms interface design

Remember to give appropriate names to the design elements of the form. This example has used the names shown in green on Figure 3.3.

The code to achieve this solution should be run under the button click event.

```
Public Class Form1
    Private Sub BTNMultiply _ Click(sender As Object,
    e As EventArgs) Handles BTNMultiply.Click
        'Declaration and initialising of required local
        variables
        Dim Number1 As Integer = 0
        Dim Number2 As Integer = 0
        Dim Multi As Integer = 0


        'Storing the values input in the textboxes to the
        variables
        Number1 = TBNumber1.Text
        Number2 = TBNumber2.Text
```

> **CONTINUED**

```
        'Storing in the local variable the multiplication result
        Multi = Number1 * Number2

        'Displaying the value held in the variable Multi in the
        output text box
        TBOutput.Text = Multi

    End Sub
End Class
```

Questions

1   Extend this program to include addition, subtraction and division buttons. It will have to be programmed using a Windows Forms application.

    Points for discussion:

    a   Should all the variables be declared locally?

    b   Is Integer an appropriate data type for all the resultant output variables?

3.9 Commenting code

You may have noticed the pseudocode and programming code for the Multiply Machine includes descriptions. For pseudocode these descriptions are started with the // symbol. For the program code the descriptions are shown in green and the line always begins with a ' symbol. These are known as **comments**. When you wish to write a comment, begin the line with a ' symbol or a // symbol if writing pseudocode.

Commenting is an important part of writing an algorithm. It will help other programmers to understand how your code works. Comments can also be very helpful to remind you how the code works when you look at code you wrote weeks or months earlier.

The computer ignores comments when the code is executed. Comments will not affect the way the code operates.

Another way you can use comments is to 'comment out' code rather than delete it when designing algorithms. You may have a working solution to one part of your system but believe there is a way to improve the already working code. Testing the new code while the previous code is still in place is impossible. Rather than delete the previous code, you can 'comment out' the previous code by placing a ' symbol at the start of each line of the previous code.

The Visual Studio 2019 Community IDE provides a way to comment many lines at once. The icon can be accessed from the main design menu (Figure 3.4, with details of comment buttons in Table 3.4).

> **KEY WORD**
>
> **commenting:** a description of the algorithm written within the code. The comments are intended to help explain how the code works. Comments are ignored by the computer when the code is executed.

3 Variables and arithmetic operators

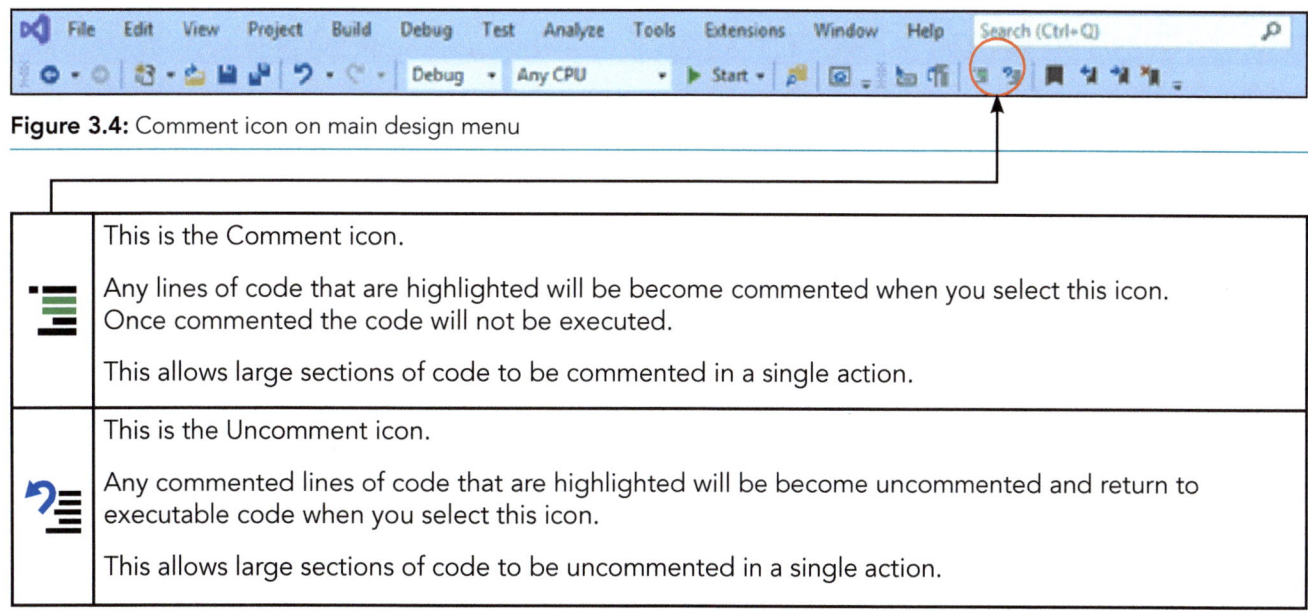

Figure 3.4: Comment icon on main design menu

≡	This is the Comment icon. Any lines of code that are highlighted will be become commented when you select this icon. Once commented the code will not be executed. This allows large sections of code to be commented in a single action.
⤺≡	This is the Uncomment icon. Any commented lines of code that are highlighted will be become uncommented and return to executable code when you select this icon. This allows large sections of code to be uncommented in a single action.

Table 3.4: Comment buttons in the IDE

3.10 Making real-world scenario-based systems

The code for the Multiply Machine is the basis of many simple real-world systems. The systems include the following steps:

- getting values from the user
- completing some form of mathematical calculation on those values (in the Multiply Machine it was a multiplication calculation)
- outputting the result of that calculation.

Examples of these real-world scenarios include:

- calculation of area or volume, such as the 'Volume of Water in an Aquarium' task
- calculation of costs and profits such as the Cinema Cost and the Sacks of Grain tasks.

Answer the questions in the practice tasks based on the demo tasks.

> ### DEMO TASK 3.2
>
> **Volume of water in an aquarium**
>
> *Design a program where the inputs will be the height, width and depth of an aquarium. The output should be the number of litres of water that the aquarium will hold (1 litre = 1000 cm^3).*

CONTINUED

Solution

Consider the series of steps to achieve this task:

1. The program will need variables to hold the three values input by the user.

    It is possible that the measurements of the aquarium will not be in exact centimetres. Users might input decimal values so will need to use REAL or DECIMAL data type for the variables.

2. The program will need to store the values input into the appropriate variable.

3. The program will need to calculate the volume of the aquarium in cubic centimetres (cm^3).

4. The program will need to calculate the number of litres the aquarium will hold by dividing the total volume by 1000.

    A constant could be used to hold the value of 1000 (the conversion factor from cm^3 to litres).

5. The program will need to display the number of litres.

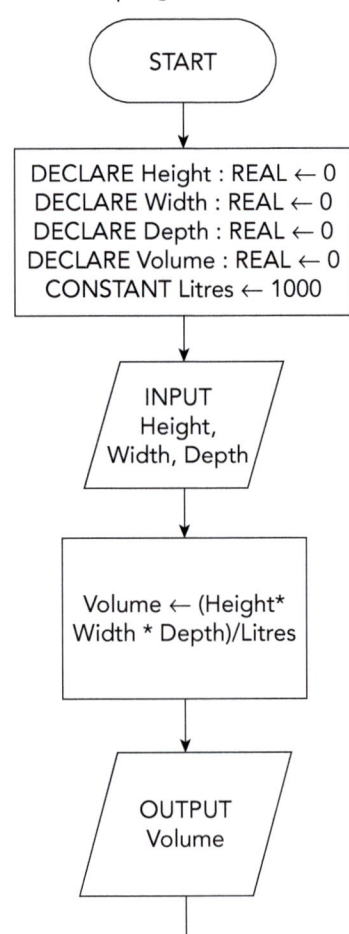

Flowchart 3.2 shows the aquarium algorithm and here is the pseudocode.

```
DECLARE Height : REAL ← 0
DECLARE Width  : REAL ← 0
DECLARE Depth  : REAL ← 0
DECLARE Volume : REAL ← 0
CONSTANT Litres ← 1000
INPUT Height, Width, Depth
Volume ← (Height*Width*Depth)/Litres
OUTPUT Volume
```

Flowchart 3.2: Flowchart for the aquarium algorithm

3 Variables and arithmetic operators

> **CONTINUED**
>
> The following code shows the console application program implementing this solution. NOTE: Observe how the logical sequence of the code follows the flowchart or pseudocode design (Flowchart 3.2).
>
> ```vb
> Module Module1
>
>     'Global variables to hold the inputs
>     'Note the addition of the 1 to the names to overcome the reserved word
>     conflict
>     Dim Height1 As Decimal = 0
>
>     Dim Width1 As Decimal = 0
>     Dim Depth1 As Decimal = 0
>
>
>     'Constant to hold the ratio of cubic centimetres to litres
>     Const Litres As Integer = 1000
>
>
>     Sub Main()
>         'Local variable to hold the resultant volume
>         Dim Volume As Decimal = 0
>
>         'Request and store user inputs
>         Console.WriteLine("Please insert Height and press Return")
>         Height1 = Console.ReadLine()
>         Console.WriteLine("Please insert Width and press Return")
>         Width1 = Console.ReadLine()
>         Console.WriteLine("Please insert Depth and press Return")
>         Depth1 = Console.ReadLine()
>
>         'Calculate Volume and store in local variable
>         Volume = Height1 * Width1 * Depth1
>         'Convert the value in volume current in cubic centimetres to litres
>         Volume = Volume / Litres
>
>         'Display the value held in the variable Volume
>         Console.WriteLine("The volume is ")
>         Console.WriteLine(Volume)
>
>         Console.ReadKey()
>     End Sub
> End Module
> ```
>
> **Note:** Windows Forms application is optional content. It is not covered in the syllabuses.

CONTINUED

Using the Windows Forms application, the interface could look something like Figure 3.5.

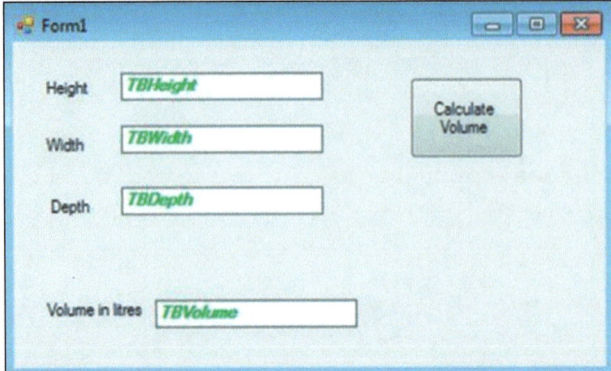

Figure 3.5: Interface design for Windows Forms aquarium algorithm

Remember to give the design elements of the form appropriate names. Figure 3.5 shows the names in green.

The Windows Forms application code to achieve this solution should be run under the button click event:

```vb
Public Class Form1

    'Global variables to hold the inputs
    'Note the addition of the 1 to the names to overcome the restricted word conflict
    Dim Height1 As Decimal = 0
    Dim Width1 As Decimal = 0
    Dim Depth1 As Decimal = 0
    'Constant to hold the ratio of cubic centimetres to litres
    Const Litres As Integer = 1000


    Private Sub BTNVolume _ Click(sender As Object, e As EventArgs) Handles BTNVolume.Click
        'Local variable to hold the resultant volume
        Dim Volume As Decimal = 0


        'placing values from the textbox into variables
        Height1 = TBHeight.Text
        Width1 = TBWidth.Text
        Depth1 = TBDepth.Text


        'Calculate Volume and store in local variable
        Volume = Height1 * Width1 * Depth1
        'Convert the value in volume current in cubic centimetres to litres
        Volume = Volume / Litres
```

CONTINUED

```
        'Output the value in local variable to the textbox
        TBVolume.Text = Volume


        'The volume calculation could have been achieved in one calculation
        'Volume = (Height1 * Width1 * Depth1)/Litres

    End Sub
End Class
```

PRACTICE TASK 3.2

Surface area and volume of a sphere

A system takes the radius (the distance from centre point to the surface) of a sphere as an input. It calculates the surface area using $A = 4\pi r^2$ and the volume of a sphere using $V = \frac{4}{3}\pi r^3$.

Draw a flowchart and create a pseudocode algorithm that will output the surface area and volume of the sphere based on the radius value input.

Test that your algorithm works by programming and running the code in Visual Basic.

DEMO TASKS 3.3–3.4

3.3 Cinema costs

A cinema has a standard admission cost of $10.00 to watch a film.

- *Adults (anyone aged between 16 and 65) are charged the standard cost*
- *Children (under the age of 16) receive a 25% reduction on the standard cost*
- *Senior citizens (people over the age of 65) receive a 10% reduction on the standard cost.*

The cinema wants a system that takes as inputs the number of Adults, Children under 16 and Senior citizens that buy tickets to watch a film. The system will output the total income (the total amount of money the cinema makes) from selling tickets for the film.

Solution

Consider the series of steps needed to achieve this task:

1. The program will need variables to hold the three values input by the user and the results of the calculations.

   People can only be counted in whole numbers (you cannot get ½ a person) so INTEGER would be the most appropriate data type for those variables.

   The results could contain decimal values so a REAL or DECIMAL data type would be the most appropriate for those variables.

2. The program will need to store the values input into the appropriate variable.

CONTINUED

3   The program will need to calculate the income from each age group.

   • The cost for each age group can be calculated from the standard $10 admission cost (Children under 16 = $10 * 0.75)

   • A constant could be used to hold the $10 standard admission cost.

4   The program will need to calculate the total income by adding the income from each age group.

5   The program will need to display the total income.

Flowchart 3.3 shows the cinema cost algorithm.

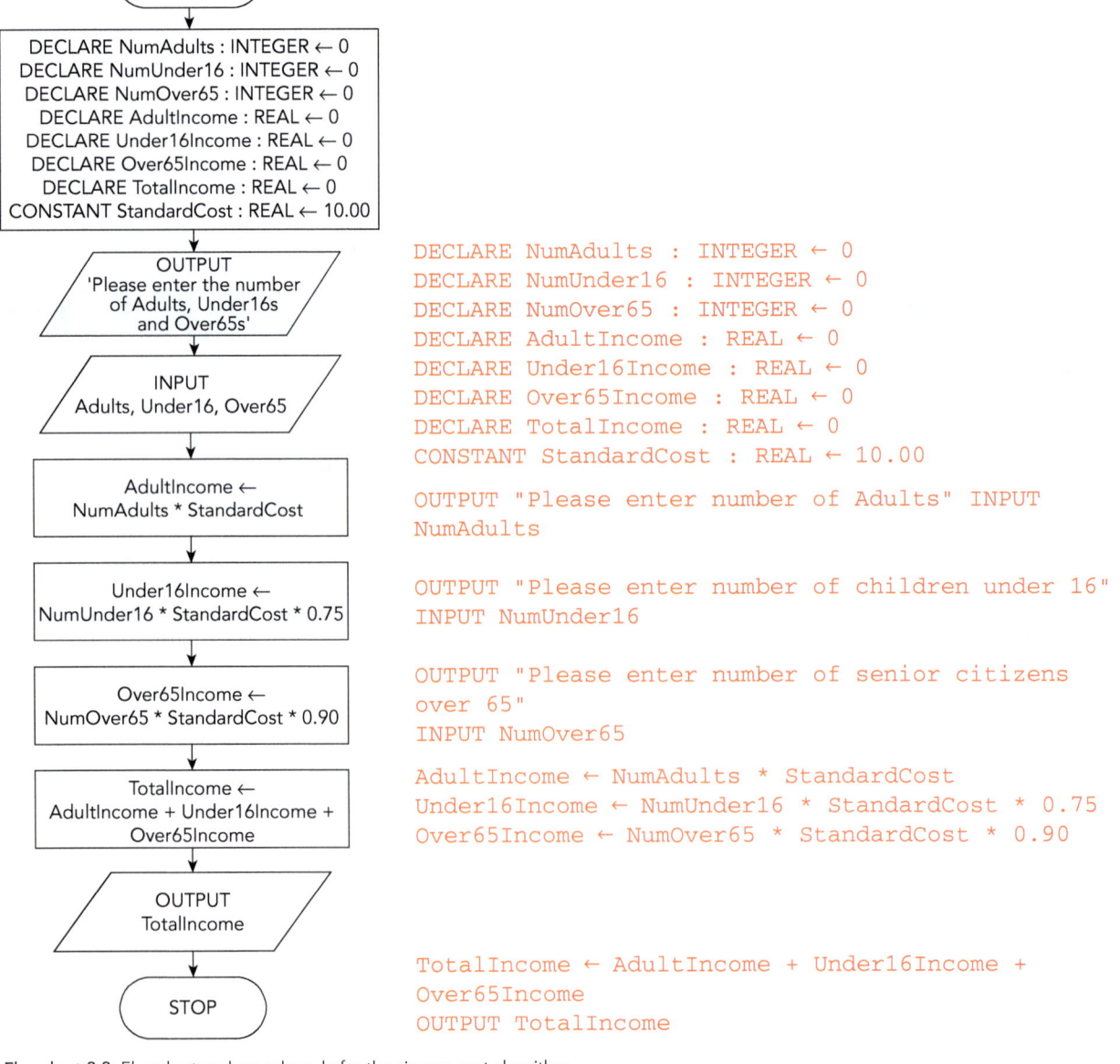

Flowchart 3.3: Flowchart and pseudocode for the cinema cost algorithm

CONTINUED

The following code shows the console application program implementing this solution:

```vb
Module Module1
    'Global Integer variables to hold the number of
    'people at the film from each age group
    Dim NumAdults As Integer = 0
    Dim NumUnder16 As Integer = 0
    Dim NumOver65 As Integer = 0
    'Global variables to hold the total cost for each group
    Dim AdultIncome As Decimal = 0
    Dim Under16Income As Decimal = 0
    Dim Over65Income As Decimal = 0
    Dim TotalIncome As Decimal = 0
    'Constant to hold standard cost
    Const StandardCost As Decimal = 10.0

    Sub Main()
        'Prompt for input from user and store values entered
        Console.WriteLine("Please enter number of Adults")
        NumAdults = Console.ReadLine()
        Console.WriteLine("Please enter number of Under 16s")
        NumUnder16 = Console.ReadLine()
        Console.WriteLine("Please enter number of Over65s")
        NumOver65 = Console.ReadLine()


        'Calculate income for individual age groups
        AdultIncome = NumAdults * StandardCost
        Under16Income = NumUnder16 * StandardCost * 0.75
        Over65Income = NumOver65 * StandardCost * 0.9
        'Calculate total income
        TotalIncome = AdultIncome + Under16Income + Over65Income
        'Output the total income

        Console.WriteLine(TotalIncome)
        Console.ReadKey()

    End Sub
End Module
```

> **CONTINUED**
>
> ### 3.4 Income from sacks of grain
>
> *Design a system for a farmer who grows grain. He harvests the grain and stores it in sacks which hold exactly 25 kg. The sacks of grain are sold and the farmer is paid in US dollars for each sack.*
>
> *The farmer wants a system that takes as an input:*
>
> - the total weight of grain collected from a field
> - the amount of money in US dollars and cents that the farmer can get for each full sack.
>
> *The system will output the number of sacks that can be filled from the grain collected and the total money the farmer will receive if he sells all the sacks.*
>
> #### Solution
>
> Consider the series of steps to achieve this task.
>
> 1. The program will need variables to hold the values input by the user.
> 2. The program will need to store the values input into the appropriate variable.
> 3. The program will need to calculate how many full sacks can be obtained from the weight of grain input.
>
>   A simple division will not give the correct answer. We will need to use an integer division and output the quotient result.
>
> 4. The program will have to calculate how much money can be made from the sacks.
>
>   Multiply the number of sacks by the amount of money each sack is worth.
>
>   As currency is a decimal value, we will not be able to use an Integer data type for the variable.
>
> 5. The program will need to display the number of sacks and total monetary value of those sacks.
>
> Flowchart 3.4 shows the flowchart for the cinema cost algorithm.

3 Variables and arithmetic operators

CONTINUED

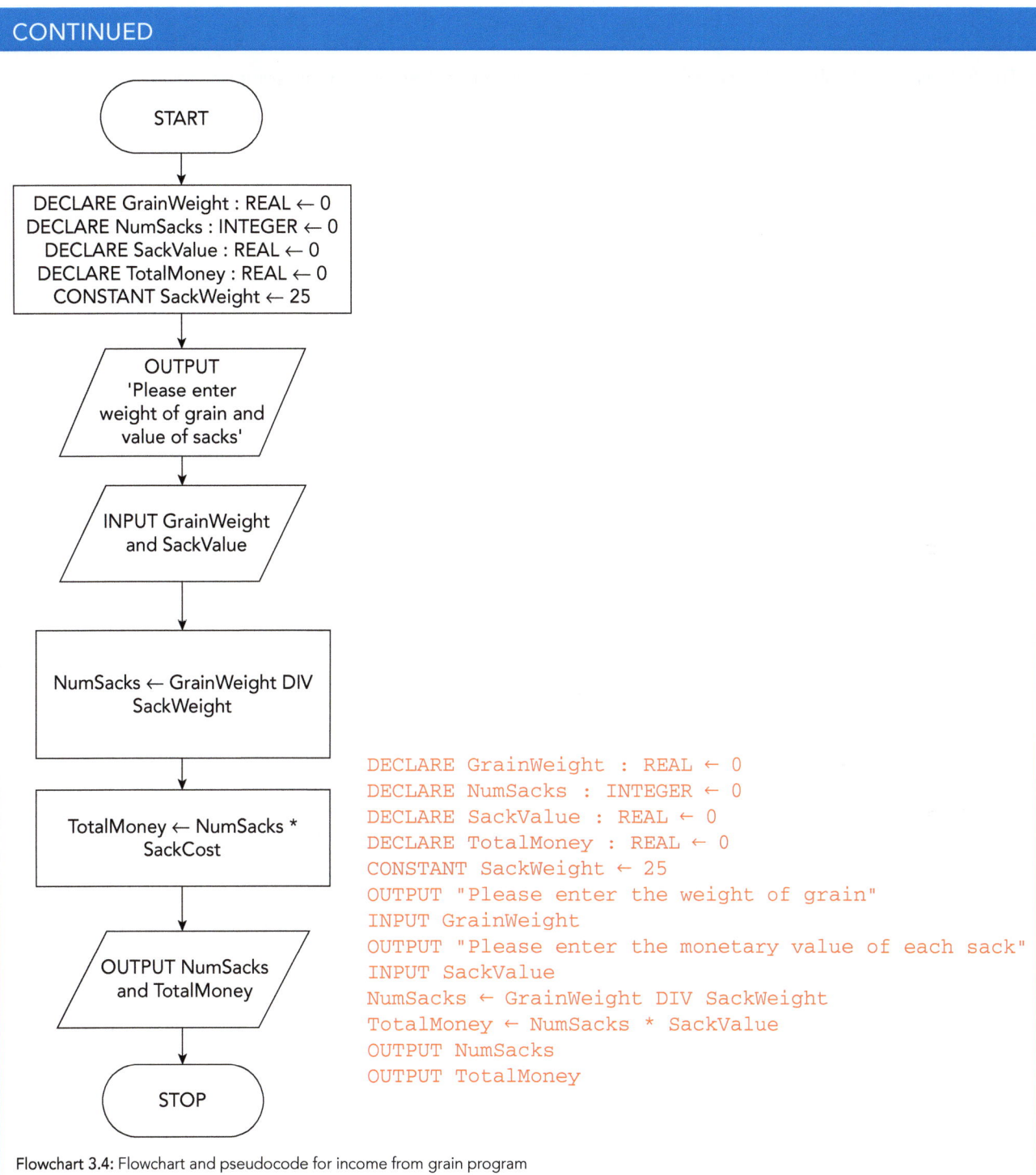

Flowchart 3.4: Flowchart and pseudocode for income from grain program

> **CONTINUED**
>
> The following code shows the console application program implementing this solution:
>
> ```vb
> Module Module1
>     'Global variables to hold input and calculated values
>     Dim GrainWeight As Decimal = 0
>     Dim SackValue As Decimal = 0
>     Dim TotalMoney As Decimal = 0
>     Dim NumSacks As Integer = 0
>     Const SackWeight = 25
>
>     Sub Main()
>         'Prompt for user input and store values entered
>         Console.WriteLine("Please enter total weight of grain in kilograms")
>         GrainWeight = Console.ReadLine()
>         Console.WriteLine("Please enter the value in dollars of each sack")
>         SackValue = Console.ReadLine()
>         'Calculate the number of complete sacks that can be filled
>         NumSacks = GrainWeight \ SackWeight
>         'Calculate the total money from all sacks
>         TotalMoney = NumSacks * SackValue
>
>         'Output the total money that could be received
>         Console.WriteLine(NumSacks)
>         Console.WriteLine(TotalMoney)
>         Console.ReadKey()
>     End Sub
> End Module
> ```

> **PRACTICE TASKS 3.3–3.4**
>
> ### 3.3 Volume of water in an aquarium
>
> 1. The current system assumes the water is level with the top of the aquarium. Alter the system so that the maximum depth of the water is 5 centimetres from the top of the aquarium.
>
> 2. Alter the system so it will work for cylindrical aquariums.
>
> ### 3.4 Sacks of grain
>
> The current system makes no allowance for grain that is dropped and lost when filling the sacks. The farmer estimates that 5% of the grain harvested is lost when filling the sacks. Alter the system to allow for this loss.

> **CHALLENGE TASKS 3.1–3.2**
>
> Challenge yourself, or your classmates, to complete a programming task. The following are some examples of the type of task you might like to consider.
>
> For each challenge, you should draw a flowchart and create a pseudocode algorithm before programming and running the code in Visual Basic.
>
> **3.1** Speed conversions
>
> Program a system to convert between different measurements of speed. The system will take as an input:
>
> - a speed in kilometres per hour (kph).
>
> The system will output the equivalent speed in:
>
> - miles per hour (mph)
> - metres per second (mps).
>
> **3.2** Draining a swimming pool
>
> Program a system which will calculate how long it will take to drain a swimming pool of water. The system will take as inputs:
>
> - the total volume of water in the swimming pool (in litres)
> - the rate at which the water can be drained from the swimming pool (in litres per minute).
>
> The system will output in minutes how long it will take to completely drain the swimming pool.
>
> Extend the system to output the result in hours and minutes.

3.11 Rounding function

It is not unusual to meet a situation where a number will require rounding. For example, multiplication or division calculations involving currency values can result in a number with many decimal places.

If you attempt to find 35% of $22.50 you will get $7.875. Rounding that value to become an appropriate currency value may be required before the value can be output.

To support this type of requirement the `ROUND()` function exists in both pseudocode and Visual Basic. Table 3.5 shows how to use this function.

Function	Pseudocode or flowchart	Visual Basic code

`ROUND()`
 	
Will return a value rounded to the number of decimal places indicated. Follows the normal mathematical rounding rules.
	`ROUND (value, number of places)`
	
	`DECLARE Num : REAL ← 1.438`
	`DECLARE RoundNum : REAL ← 0`
	`RoundNum ← ROUND(Num, 2)`
	`OUTPUT RoundNum`
	
	Would output 1.44
		`Math.Round(value, number of places)`
		
		`Dim Num As Decimal = 1.438`
		`Dim RoundNum As Decimal`
		`RoundNum = Math.Round(Num, 2)`
		`Console.WriteLine(RoundNum)`
		
		Would output 1.44

Table 3.5: Rounding in Visual Basic, pseudocode and flowcharts

In addition to `ROUND()` function, Visual Basic includes functions that will round up or down to an integer value. These functions do not exist in pseudocode but rounding up and down can be achieved by using the MOD operator to identify if the fractional part of the original value is greater than or equal to 0.5. The pseudocode and Visual Basic methods are shown in Table 3.6.

Pseudocode	Visual Basic code

To Round UP
 	
`DECLARE Num : REAL ← 1.438`
`DECLARE RoundNum : REAL ← 0`
`IF Num MOD 1 >= 0.5`
`  THEN`
`    RoundNum ← ROUND(Num, 0)`
`  ELSE`
`    RoundNum ← ROUND(Num, 0)+ 1`
`ENDIF`
`OUTPUT RoundNum`
	`Math.Ceiling(value)`
	
	`Dim Num As Decimal = 1.438`
	`Dim RoundNum As Decimal`
	`RoundNum = Math.Ceiling(Num)`
	`Console.WriteLine(RoundNum)`
	
	Would output 2.0

To Round DOWN
 	
`DECLARE Num : REAL ← 1.438`
`DECLARE RoundNum : REAL ← 0`
`IF Num MOD 1 < 0.5`
`  THEN`
`    RoundNum ← ROUND(Num, 0)`
`  ELSE`
`    RoundNum ← ROUND(Num, 0)- 1`
`ENDIF`
`OUTPUT RoundNum`
	`Math.Floor(value)`
	
	`Dim Num As Decimal = 1.438`
	`Dim RoundNum As Decimal`
	`RoundNum = Math.Floor(Num)`
	`Console.WriteLine(RoundNum)`
	
	Would output 1.0

Table 3.6: Rounding up and down in Visual Basic

3.12 Random function

Another function created specifically to work with numbers is the `RANDOM()` function. The function will return a randomly generated real number between 0 and 1. This value can then be used to create random numbers in any range required. If a random whole number is required the `ROUND()` function can be used to round the real number generated to an integer value.

Table 3.7 shows the pseudocode and Visual Basic approach to generating random numbers.

3 Variables and arithmetic operators

Pseudocode or flowchart	Visual Basic

`RANDOM()`
 	
The function generates a real (decimal) number between 0 and 1. To generate a number in a larger range this value would need to be multiplied by the upper limit of the required range.
 	
For example, the following pseudocode will randomly generate integer values between 0 and 200.
 	
`DECLARE RandInt : INTEGER ← 0`
`DECLARE RandNum : REAL ← 0`
`// Obtain the random value 0 to 1`
`RandNum ← RANDOM()`
`// Multiply by 200 to increase range`
`RandNum ← RandNum * 200`
`// Use ROUND() to round to an integer`
`// set decimal places to 0 to get a whole number`
`RandInt ← ROUND(RandNum, 0)`
 	
It would also be possible to write these steps as one line of code.
 	
`RandInt ← ROUND (RANDOM() * 200, 0)`
	`Rnd()`
	
	This function generates a real (decimal) number between 0 and 1. It is used in the same way as the pseudocode function.
	
	For example, the following Visual Basic code will randomly generate integer values between 0 and 200.
	
	`Dim RandInt As Integer = 0`
	`RandInt = Math.Round(Rnd() * 200, 0)`

Table 3.7: Generating random numbers in Visual Basic, pseudocode and flowcharts

SUMMARY

Programs use variables and constants to hold values.

Variables and constants can be declared as either Global or Local:

- Local variables or constants can only be accessed by the subroutine or statement in which they have been declared. Using local declarations can help avoid accidental changing of a variable by another subroutine.
- Global variables or constants are declared outside of any subroutine. These can be accessed by any subroutine or statement in the program.

Variables and constants have identifiers (names). Identifiers are used to refer to variables and constants in the program. The identifiers given to variables or constants must be meaningful. Avoid single character identifiers.

The value of a variable can be changed during the execution of the program. The values within constants cannot be changed while the program is running.

> ### CONTINUED
>
> It is important to select the appropriate data type for the variables and constants. Using inappropriate data types could result in the program crashing or producing unexpected results.
>
> Mathematical operators can be used with values held in numeric variables. The common rules of mathematical precedence (BODMAS) apply:
>
> - Addition – represented by a + symbol: 5 + 2 = 7
> - Subtraction – represented by a – symbol: 5 – 2 = 3
> - Multiplication – represented by a * symbol: 5 * 2 = 10
> - Division – represented by the / symbol: 5 / 2 = 2.5
> - Power of – represented by the ^ symbol: 5^2 = 25
> - Modulus (Integer division remainder) – represented by the term MOD: 5 MOD 2 = 1
> - Quotient (Integer division) – represented by the \ symbol: 5 \ 2 = 2.
> - Pseudocode for quotient division uses the term DIV: 5 DIV 2 = 2
>
> Integer division can be useful when calculating ratios or finding reminders.
>
> When designing algorithms, it is crucial to consider the logical sequence of execution. It is important to declare and initialise appropriate variables as well as obtaining user input before completing any processing.
>
> Commenting code to explain how the code works is an important aspect of good practice. Comments help people who read your code to understand how it works. Computers ignore comments when the code is executed.

END-OF-CHAPTER TASKS

For each task, you should draw a flowchart and create a pseudocode algorithm before programming and running the code in Visual Basic.

1. A system is required to calculate the cost of a taxi journey. The cost is calculated by the formula:
   - a fixed cost of $5.00 for the first 2 miles
   - plus $1.50 per mile for any additional miles travelled.

   Produce a system that will take the total distance travelled as an input and output the total cost of the journey.

2. A web design company contacts potential customers by phone to try and sell them a new website. The sales personnel receive a weekly income calculated by the formula:
   - a salary of $15.50 for each hour worked in the week
   - plus a commission of $50.00 for any website sold that week.

   Once the total weekly income has been calculated, 20% of the total value is deducted for tax and then the balance is paid to the salesperson.

   The web design company need a system to calculate the income for each salesperson.

   Produce a system that will take the number of hours the salesperson has worked and the number of sales they have made in a week as inputs. The system should output the amount of tax deducted and the remaining income after the tax deduction.

Chapter 4
Selection

IN THIS CHAPTER YOU WILL:

- understand how selection can be used to allow a program to follow different paths of execution
- understand how selection is shown in flowcharts and pseudocode
- understand the differences between and the advantages of using:
    - IF...ELSE...ENDIF statements
    - nested IF statements
    - IF...ELSEIF...ELSE...ENDIF statements
    - CASE...OTHERWISE...ENDCASE statements
- understand how to use logical operators when programming selection algorithms.

Introduction

Making decisions is a part of everyday life. To make a decision, we consider the factors that influence that decision. For example, when teachers decide what grade to give to an assignment, they would normally work out how many marks the student has achieved and compare this with the grade boundaries for that assignment (for example, 60–69% would be a B, 70% and above would be an A).

Computers can be programmed to make decisions in a similar way. Using **selection** allows a programmer to write a program that can follow different paths through the execution of the code. The paths the code follows are determined by **criteria**, the electronic equivalent of the factors that influence human decisions. Simple decisions, such as awarding an assignment grade, can be easily replicated in a program. More complex programs will involve a much greater number of criteria.

4.1 The need for selection

Systems often need to be programmed to complete different processes depending on the input. For example, an automatic door will open if it detects that someone wishes to enter or it will remain shut when no presence is detected. Expert systems will provide answers or conclusions based on the user response to previous questions. Both of these systems appear to be able to make decisions. However, the reality is that the systems have been logically designed to complete a certain process based on the expected input.

Visual Basic and many other languages achieve this by using programming techniques known as **IF statements** or **CASE statements**. Both techniques perform a similar role, although CASE statements are generally considered more appropriate for situations where many decision criteria need to be considered.

4.2 IF statements

If the logic process for control of an automatic door was written down it might appear as, 'If a presence is detected then open the door, otherwise close it'. This is very similar to coding an IF statement:

1. The code provides a **condition** to evaluate ('if a presence is detected'); the outcome of the condition is either True or False.
2. The code then provides actions depending on the outcome of the condition (if it is True 'open the door'; if it is False 'close the door').

It is common to use flowcharts and pseudocode when designing algorithms that use IF statements. In a flowchart, the symbol used to indicate a decision is a diamond. The diamond contains information about the criteria and normally has two exit routes indicating the TRUE and FALSE paths.

Flowchart 4.1 includes the decision symbol. The TRUE and FALSE paths have been indicated. Once the appropriate action has been performed, the program flow returns to 'Check for Presence' and the input is again evaluated by the IF statement.

> **KEY WORDS**
>
> **selection:** code follows a different sequence based on what condition is chosen.
>
> **criteria:** the specific rules that are used by the program to recognise if a condition is TRUE or FALSE. The singular form of criteria is criterion.
>
> **IF statement:** a statement that allows a program to follow or ignore a sequence of code depending on a Boolean condition.
>
> **CASE statement:** a simple method of providing multiple paths through the code based on a single variable or user input.
>
> **condition:** the criteria that are tested as part of the execution of the code. The condition will result in either a True or False answer when tested.

The format for the pseudocode of an IF statement is very close to the actual Visual Basic code for an IF statement:

```
IF condition to test
  THEN
    //code to follow if condition is TRUE
  ELSE
    //code to follow if condition is FALSE
ENDIF
```

We can see this format in the pseudocode for the automatic door algorithm:

```
IF Presence = True
  THEN
    Open Door
  ELSE
    Close door
ENDIF
```

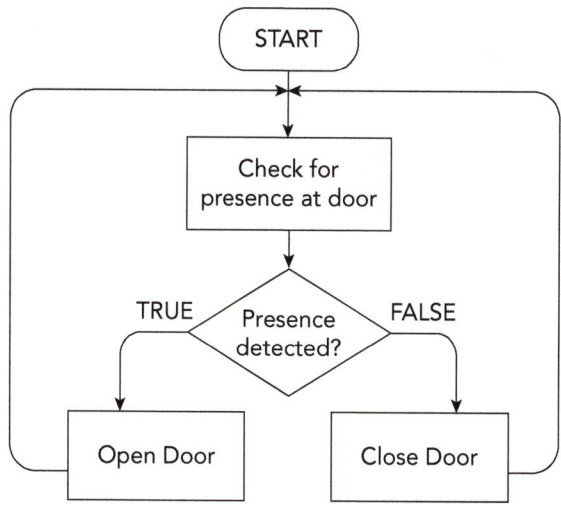

Flowchart 4.1: Flowchart for the automatic door

This is, of course, an oversimplification of the logic of an automatic door. It does not provide for the door to be set so that it is constantly open. Also, a real door would be constantly checking for a presence, it would not complete the door close process before checking.

4.3 Single action IF statements

The ELSE in an IF statement is a very useful feature as it provides a single command method of indicating an alternative code path to follow in all situations where the criteria are not TRUE.

However, there are situations where a decision does not have an alternative action – the alternative action is simply to do nothing. Similarly, not all IF statements will have an alternative action and therefore the ELSE may be omitted. This would be appropriate in a situation where the decision results in only one action.

For example, a system used to calculate the cost of a train fare could apply a discount if the passenger is a child. The system would first calculate the normal fare and then only apply the discount if the passenger was a child. If the passenger were an adult, no additional action would be taken. This pseudocode is for that algorithm:

```
Calculate TicketPrice
IF Passenger is Child
  THEN
    TicketPrice ← TicketPrice * Discount
ENDIF
```

4.4 Logical operators

In the ticket price example, the criteria was 'Passenger is Child'. Real-life systems would usually be more precise in creating a criterion to determine if a passenger was a child. It would probably be based around age – for example, 'passenger is under 16 years of age'. Other systems will also require precise criteria. An air-conditioning system will receive continuous temperature data and will perform actions based on precise temperature values. A system for determining examination grades will calculate the grade output by identifying if the students' marks fall within certain grade boundaries.

For these types of decisions, a number of logical operators exist. The basic operators supported by Visual Basic are shown in Table 4.1.

Operator	Description
=	Is equal to
>	Is greater than
<	Is less than
>=	Is greater than or equal to
<=	Is less than or equal to
<>	Is not equal to

Table 4.1: Logical operators

The choice of the correct logical operator is important. Using the wrong operator can produce unexpected results in your algorithms. Often the way in which the decision to be made is worded will indicate the appropriate operator to use. Table 4.2 shows some applications of logical operators and some common errors.

Decision in Words	Appropriate Operator	Common Errors
Apply a discount for students aged under 16.	`IF Student < 16 THEN`	• **Using >** `IF Student > 16` would apply the discount for students over 16. • **Using <=** `IF Student <= 16` would also apply the discount for students aged 16. The wording states UNDER.
Turn on the cooler when the temperature is 10°C or more.	`IF Temp >= 10 THEN`	• **Using =** `IF Temp = 10` means the cooler will only operate when the temperature is exactly 10°C. If the temperature rises above 10°C the condition would no longer be true and the cooler would stop. • **Using >** `IF Temp > 10` means the cooler would not turn on at 10°C as required.

Table 4.2: Example choices for logical operators

4.5 Coding IF statements in Visual Basic

The code for an IF statement in Visual Basic is very similar to the pseudocode version. Remember, one advantage of using Visual Basic is the similarity between the actual programming code and pseudocode.

This is an example of Visual Basic code to determine if the age input by the user is less than 16:

```
Dim Age As Integer
Console.WriteLine("Enter Age in whole years")
Age = Console.ReadLine()
If Age < 16 Then
    'code to execute if condition TRUE
Else
    'code to execute of condition FALSE
End If
```

> **TIP**
>
> When typing an If statement, the IDE software will automatically apply the `End If` once you complete the `If ... Then` and press the Return key. Should an `Else` be required you will need to type this. The IDE will automatically indent all the code within the statement.

DEMO TASK 4.1

A system is required that will take two whole numbers as input. If the second number is larger than the first, the system will output 'Second', if not the output will be 'First'.

Solution

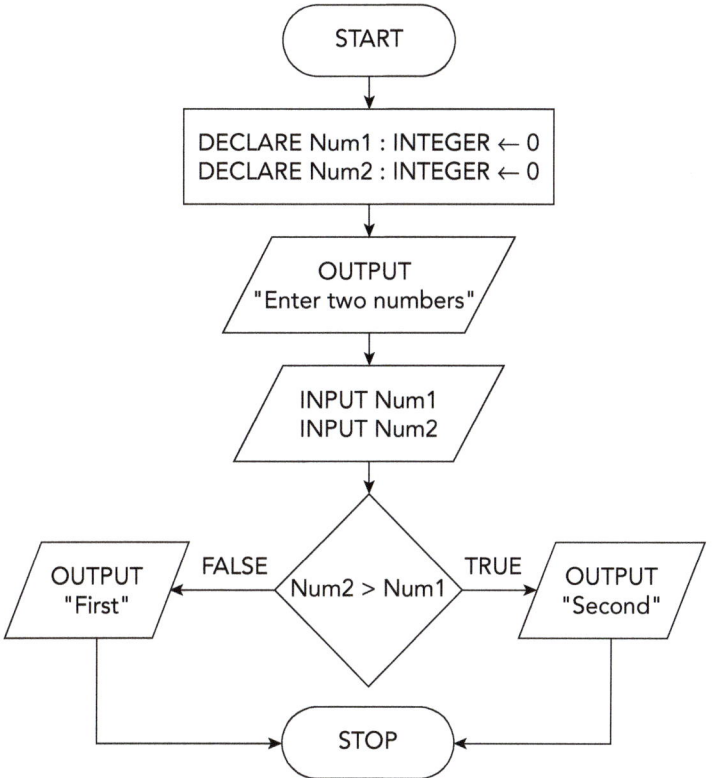

Flowchart 4.2 shows solution. Here is the pseudocode solution:

```
DECLARE Num1 : INTEGER ← 0
DECLARE Num2 : INTEGER ← 0
OUTPUT "Enter two numbers"
INPUT Num1
INPUT Num2
IF Num2 > Num1
   THEN
      OUTPUT "Second"
   ELSE
      OUTPUT "First"
ENDIF
```

Flowchart 4.2: Flowchart solution

The code for a console application implementation of this algorithm is shown below:

```
Sub main()
    Dim Num1 As Integer = 0
    Dim Num2 As Integer = 0
    Console.WriteLine("Enter first number")
    Num1 = Console.ReadLine()
    Console.WriteLine("Enter second number")
    Num2 = Console.ReadLine()

    If Num2 > Num1 Then
        Console.WriteLine("Second")

    Else
        Console.WriteLine("First")
    End If
    Console.ReadKey()
End Sub
```

> **CAMBRIDGE IGCSE™ & O LEVEL COMPUTER SCIENCE: PROGRAMMING BOOK**

> **CONTINUED**
>
> 1. What do you think will happen if the same number is input twice?
>
> 2. If the question is re-worded to read, 'If the second number is the same as or larger than the first the system will output "Second", if not the output is "First".', how would you change the algorithm to meet this change?

> **PRACTICE TASKS 4.1–4.3**
>
> **4.1a** Design a flowchart and pseudocode of an algorithm for a system that takes two numbers. If the numbers are the same, the system outputs 'Match'. If the numbers are different, the system outputs 'Not Matched'.
>
> **b** Implement and test your algorithm in a Visual Basic console application or Windows Forms application.
>
> **4.2a** Design a flowchart and pseudocode of an algorithm for a system that takes two integer values. If the second number is more than twice the value of the first number, the system outputs 'Too Large', otherwise the system outputs 'Acceptable'.
>
> **b** Implement and test your algorithm in a Visual Basic console application or Windows Forms application.
>
> **4.3a** Design a flowchart and pseudocode of an algorithm for a system that takes two integer values. If the second number is a factor of the first number, the system outputs 'Factor', otherwise the system outputs 'Not a Factor'.
>
> **b** Implement and test your algorithm in a Visual Basic console application or Windows Forms application.

> **TIP**
>
> A factor of a number is any value that will divide exactly into that number. For example, the factors of 30 are 1, 2, 3, 5, 6, 10, 15 and 30.

4.6 Multiple decisions

The IF statements discussed so far have included only one criterion and a maximum of two paths. We could change the previous example to output one of three possible values:

- 'Second' if the second number is larger than the first.
- 'First' if the second number is smaller than the first.
- 'The Same' if the two numbers are equal.

Now we have three decisions to make.

This is an example of a scenario that involves multiple decisions. It is possible to solve this type of a programming challenge in different ways. This section identifies the different approaches that can be used and the advantages and disadvantages of those approaches.

Sequential IF statements

Sequential IF statements are a series of IF statements that follow each other. Each statement checks just one criterion and ends before the next one starts.

Pseudocode

```
DECLARE Num1 : INTEGER ← 0
DECLARE Num2 : INTEGER ← 0
OUTPUT "Enter first number"
INPUT Num1
OUTPUT "Enter second number"
INPUT Num2

IF Num2 > Num1
  THEN
    OUTPUT "Second"
ENDIF
IF Num2 < Num1
  THEN
    OUTPUT "First"
ENDIF
IF Num2 = Num1
  THEN
    OUTPUT "The same"
ENDIF
```

Visual Basic code

```
Sub main()
    Dim Num1 As Integer = 0
    Dim Num2 As Integer = 0
    Console.WriteLine("Enter first number")
    Num1 = Console.ReadLine()
    Console.WriteLine("Enter second number")
    Num2 = Console.ReadLine()

    If Num2 > Num1 Then
        Console.WriteLine("Second")
    End If

    If Num2 < Num1 Then
        Console.WriteLine("First")
    End If

    If Num2 = Num1 Then
        Console.WriteLine("The same")
    End If
    Console.ReadKey()
End Sub
```

Code snippet 4.1: Pseudocode and Visual Basic console application implementation of sequential IF statements

Although this approach achieves the required outcome, it is inefficient and is not considered an effective solution.

Consider the situation where the second number is larger than the first. The first IF statement in the sequence will have a TRUE condition and provide the appropriate output. This means the condition of the remaining IF statements must be FALSE. But the code must still execute the remaining IF statements even if nothing is output to the screen. The algorithm produces the required output, but two IF statements have been executed unnecessarily.

Nested IF statements

To avoid the inefficiency of multiple IF statements, it is possible to place one or more IF statements entirely within another. Each of the following IF statements will be executed only if the first condition prove to be false. These are known as **nested IF statements**.

Code snippet 4.2 shows, in pseudocode and VB code, how a nested IF approach could be applied to the inefficient sequence of IF statements shown in Code snippet 4.1. Because the second IF statement will only execute if the criteria in the first statement is False, unnecessary execution of IF statements is avoided.

> **KEY WORD**
>
> **nested IF statement:** an IF statement with the ability for additional conditions to be checked once earlier conditions have determined a path.

Pseudocode

```
DECLARE Num1 : INTEGER ← 0
DECLARE Num2 : INTEGER ← 0
OUTPUT "Enter first number"
INPUT Num1
OUTPUT "Enter second number"
INPUT Num2

IF Num2 > Num1
  THEN
    OUTPUT "Second"
  ELSE
    IF Num2 < Num1
      THEN
        OUTPUT "First"
      ELSE
        OUTPUT "The same"
    ENDIF
ENDIF
```

Visual Basic code

```
Sub main()
    Dim Num1 As Integer = 0
    Dim Num2 As Integer = 0
    Console.WriteLine("Enter first number")
    Num1 = Console.ReadLine()
    Console.WriteLine("Enter second number")
    Num2 = Console.ReadLine()

    If Num2 > Num1 Then
        Console.WriteLine("Second")
    Else
        If Num2 < Num1 Then
            Console.WriteLine("First")
        Else
            Console.WriteLine("The same")
        End If
    End If
    Console.ReadKey()
End Sub
```

Code snippet 4.2: Pseudocode and Visual Basic console application implementation of nested IF statements

The pseudocode and actual Visual Basic code are very similar. NOTE: Observe how the indentation improves the readability of the code. When writing pseudocode remember to indent your code.

Nested IF statements are shown in flowcharts by chaining the decision symbols (see Flowchart 4.3).

The FALSE path of one decision leads to the input of another decision.

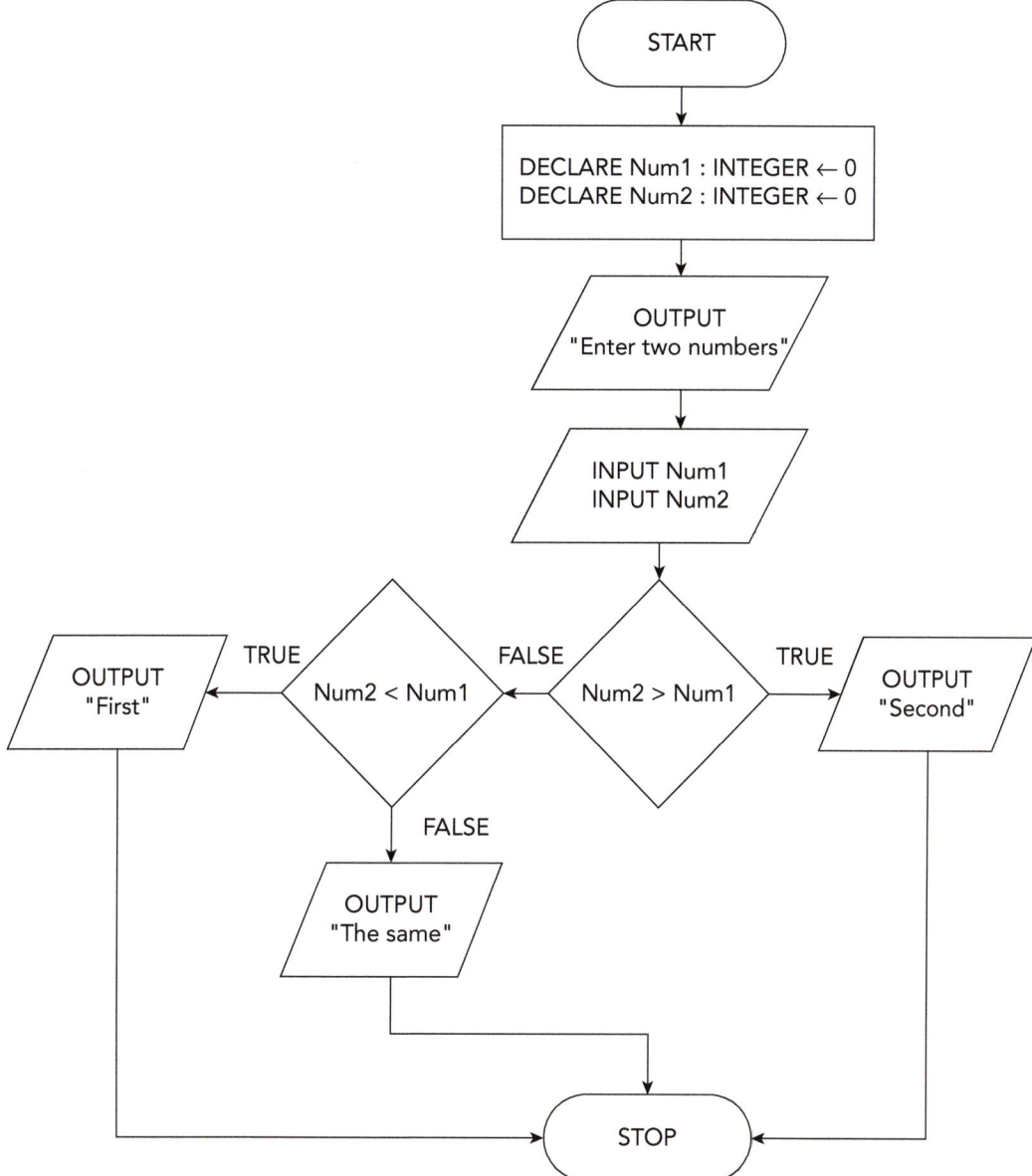

Flowchart 4.3: Flowchart to show nested IF statement

ELSEIF statements

In Visual Basic, and some other languages, it is possible to write the equivalent of a series of nested IF statements by making use of an **ELSEIF statement**. It is executed in a similar way but is easier to read.

The format for showing these statements in flowcharts is the same as the format for showing nested IF statements – a series of chained decision symbols. But the format for both pseudocode and actual programming code is different from the nested IF statements.

> **KEY WORD**
>
> **ELSEIF statement:** an alternative to nested IF statements. Each condition is tested in turn. When a condition is TRUE, the code related to that condition is executed and the statement is ended.

```
If 'first criteria' Then
    'Code to run if first
    criteria is TRUE
ElseIf 'second criteria' Then
    'Code to run if second
    criteria is TRUE
Else
    'Code to run if all other
    criteria are FALSE
End If
```

The execution of this statement follows a specific path.

- The first criteria is checked – if this is TRUE, the relevant code is executed and the statement ends and no further criteria are checked. IF the first criteria is FALSE, the statement will then check the second criteria.

- The second criteria is checked – if this is TRUE, the relevant code is executed and the statement ends without checking any further criteria.

- Only when all criteria have been checked and found to be FALSE would the statement execute the code within the ELSE part of the code.

The pseudocode and programming code for using this approach to the two numbers task are shown in Code snippet 4.3.

Pseudocode

```
DECLARE Num1 : INTEGER ← 0
DECLARE Num2 : INTEGER ← 0

OUTPUT "Enter first number"
INPUT Num1
OUTPUT "Enter second number"
INPUT Num2

IF Num2 > Num1
  THEN
    OUTPUT "Second"
ELSEIF Num2 < Num1
  THEN
    OUTPUT "First"
  ELSE
    OUTPUT "The same"
ENDIF
```

Visual Basic code

```
Sub main()
    Dim Num1 As Integer = 0
    Dim Num2 As Integer = 0
    Console.WriteLine("Enter first number")
    Num1 = Console.ReadLine()
    Console.WriteLine("Enter second number")
    Num2 = Console.ReadLine()

    If Num2 > Num1 Then
        Console.WriteLine("Second")
    ElseIf Num2 < Num1 Then
        Console.WriteLine("First")
    Else
        Console.WriteLine("The same")
    End If
    Console.ReadKey()
End Sub
```

Code snippet 4.3: Pseudocode and Visual Basic console application implementation of ELSEIF statement

> **TIP**
>
> Using multiple ELSEIF elements can allow for multiple conditions to be checked in a single IF statement.
>
> Consider this code where a system requires a user to input DIVIDE, MODULUS, QUOTIENT to select the correct division mathematical process to execute.
>
> ```
> If Choice = "DIVIDE" Then
>     Console.WriteLine(Num1 / Num2)
> ElseIf Choice = "MODULUS" Then
>     Console.WriteLine(Num1 Mod Num2)
> ElseIf Choice = "QUOTIENT" Then
>     Console.WriteLine(Num1 \ Num2)
> Else
>     Console.WriteLine("Not a valid Choice")
> End If
> ```
>
> The final ELSE can be used to identify incorrect entries.

Working with scale of values

When using nested IF statements or ELSEIF statements to check on a scale of values, it is crucial that the criteria are provided in the correct order.

> ### PRACTICE TASKS 4.4–4.5
>
> A system is required to calculate and output an examination grade using these grade boundaries:
>
> | Grade | Boundary |
> |---|---|
> | Merit | Score above 25 |
> | Pass | Score between 15 and 24 |
> | Fail | Score below 15 |
>
> Two algorithms have been created (Code snippet 4.4). One of the algorithms will not always produce the expected output values.
>
> **4.4a** Which algorithm (A or B) will generate the expected output?
>
>   **b** What input values will cause the incorrect algorithm to error?
>
> Algorithm A
> ```vb
> Dim Score As Integer
> Score = Console.ReadLine()
> If Score > 25 Then
>     Console.WriteLine("Merit")
> ElseIf Score >= 15 Then
>     Console.WriteLine("Pass")
> Else
>     Console.WriteLine("Fail")
> End If
> ```
>
> Algorithm B
> ```vb
> Dim Score As Integer
> Score = Console.ReadLine()
> If Score >= 15 Then
>     Console.WriteLine("Pass")
> ElseIf Score > 25 Then
>     Console.WriteLine("Merit")
> Else
>     Console.WriteLine("Fail")
> End If
> ```
>
> Code snippet 4.4: Algorithm A and algorithm B
>
> **4.5a** Design a flowchart and pseudocode of an algorithm for a system that calculates the cost of a train ticket. The system takes as inputs the price of the ticket and the age of the passenger. The system outputs the final cost of the ticket using the information in the table:
>
> | Age of Passenger | Discount | Final Ticket Price |
> |---|---|---|
> | Age 18 or above | No discount | Full price is charged |
> | Age 15 or above but less than 18 | 20% discount | Ticket is charged at 80% of full price |
> | Age 4 or above but less than 15 | 40% discount | Ticket is charged at 60% of the full price |
> | Aged under 4 | 100% discount | No charge, ticket is free |
>
>   **b** Using nested IF statements, implement and test your algorithm in a Visual Basic console application or Windows Forms application.
>
>   **c** Using ELSEIF statements, implement and test your algorithm in a Visual Basic console application or Windows Forms application.

> **CHALLENGE TASK 4.1**
>
> a   Design a flowchart and pseudocode of an algorithm for a system that takes as an input the name of a month and outputs the number of days in that month. Assume February will output 28 days.
>
> b   Implement and test your algorithm in a Visual Basic console application or Windows Forms application.

4.7 CASE statements

CASE statements are considered an efficient alternative to multiple IF statements or ELSEIF statements. In a CASE statement, selection is based on one variable with multiple possible values. CASE statements allow complex situations, based on a single variable, to be programmed more easily than using IF statements.

Consider the situation where a user must choose to input one of A, B or C. The code is required to follow different paths depending on which letter the user has chosen to input. The pseudocode in Code snippet 4.5 compares the approach taken using an ELSEIF statement with the approach taken using a CASE statement.

ELSEIF statement

```
IF Choice = "A"
   THEN
      //code to follow
ELSEIF Choice = "B"
   THEN
      //code to follow
ELSEIF Choice = "C"
   THEN
      //code to follow
   ELSE
      OUTPUT "Incorrect input"
ENDIF
```

CASE statement

```
CASE OF Choice
   "A" : 'code to follow'
   "B" : 'code to follow'
   "C" : 'code to follow'
   OTHERWISE OUTPUT "Incorrect input"
ENDCASE
```

Code snippet 4.5: A comparison of an ELSEIF and a CASE statement

Both approaches achieve the same outcome and are easier to read than the equivalent nested IF solution. Some programmers believe that a CASE statement is simpler to code and easier to read than the ELSEIF statement, particularly where many decision criteria are involved.

Flowchart 4.4 shows the two main approaches to drawing a flowchart for a CASE statement:

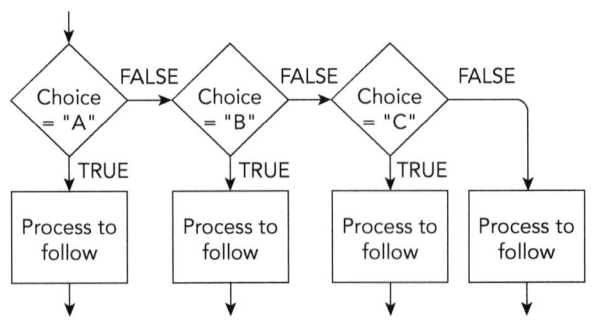

Flowchart 4.4(a): Flowchart using sequential decisions

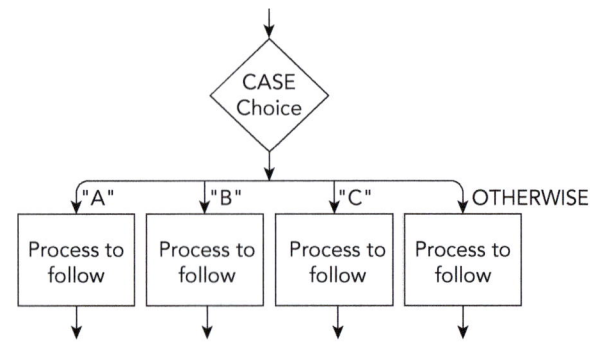

Flowchart 4.4(b): Flowchart with multiple branches

Flowchart 4.4(a) shows a sequence of connected decisions to show the logical process. This is the same flowchart that would show the nested IF or ELSEIF approach.

Flowchart 4.4(b) shows an alternative approach used by some programmers. A specific CASE symbol leads to multiple decision branches including the OTHERWISE branch.

4.8 Coding CASE statements in Visual Basic

Programming a CASE statement closely follows the pseudocode format of a CASE statement.

```
Select Case Choice
    Case "A"
        'code to follow'
    Case "B"
        'code to follow'
    Case "C"
        'code to follow'
    Case Else
        Console.WriteLine("Incorrect input")
End Select
```

However Visual Basic uses specific keywords that differ from those traditionally used in pseudocode.

The CASE statement has four main elements, described in Table 4.3.

Element	Description
`Select Case Choice`	The start of a CASE statement. The identifier (name) of the variable on which the selection is to be based follows the `Case` keyword.
`Case "A"` `'code to follow'`	A series of statements identifying the values on which the decision is based. The code path to be followed if the condition is true is indented after the statement.
`Case Else`	The path that is to be followed if all previous conditions are False. Similar to the ELSE in an IF statement, this clause does not have to be included in a CASE statement.
`End Select`	The end of the CASE statement.

Table 4.3: CASE statement elements

DEMO TASK 4.2

In Chapter 3 we created a calculator that could take two numbers and perform basic arithmetic functions on those numbers depending on which button the user pressed. A real calculator does not work in this way. Instead the first number is input, the arithmetic operator is selected, the second number is input, and the answer is obtained by pressing the 'equals' button.

The task is to create a calculator that works like a real calculator.

Solution

A flowchart design for this algorithm could look like Flowchart 4.5.

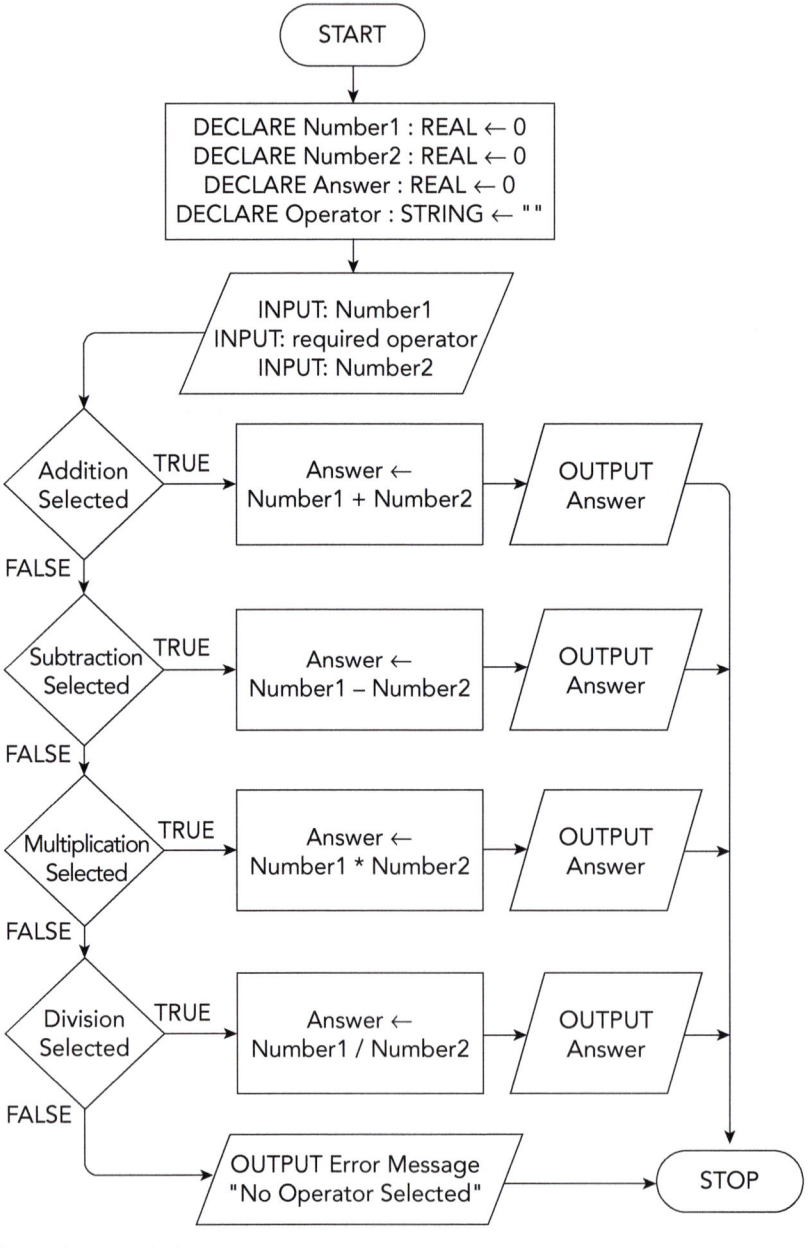

Flowchart 4.5: Flowchart for simulating a calculator

> **CONTINUED**
>
> The flowchart shows a series of connected decision symbols.
>
> Each decision has a TRUE path that completes the appropriate mathematical calculation, outputs the result of that calculation and ends the program.
>
> The FALSE path from the decisions becomes the input for the next decision.
>
> This flowchart indicates the logic of the algorithm but does not define the specific approach to be taken. The same flowchart would be appropriate for either a CASE statement or an IF statement approach.
>
> The pseudocode of the algorithm is more specific about the approach to adopt.
>
> CASE statement approach
>
> ```
> DECLARE Num1 : REAL ← 0
> DECLARE Num2 : REAL ← 0
> DECLARE Choice : STRING ← ""
> OUTPUT "Enter first number"
> INPUT Num1
> OUTPUT "Enter operation"
> OUTPUT "A = Add, S = Subtract"
> OUTPUT "M = Multiply, D = Divide"
> INPUT Choice
> OUTPUT "Enter second number"
> INPUT Num2
>
> CASE OF Choice
>     "A" : OUTPUT Num1 + Num2
>     "S" : OUTPUT Num1 - Num2
>     "M" : OUTPUT Num1 * Num2
>     "D" : OUTPUT Num1 / Num2
>     OTHERWISE
> OUTPUT "Incorrect Entry"
> ENDCASE
> ```
>
> ELSEIF statement approach
>
> ```
> DECLARE Num1 : REAL ← 0
> DECLARE Num2 : REAL ← 0
> DECLARE Choice : STRING ← ""
> OUTPUT "Enter first number"
> INPUT Num1
> OUTPUT "Enter operation"
> OUTPUT "A = Add, S = Subtract"
> OUTPUT "M = Multiply, D = Divide"
> INPUT Choice
> OUTPUT "Enter second number"
> INPUT Num2
>
> IF Choice = "A"
>    THEN
>       OUTPUT Num1 + Num2
> ELSEIF Choice = "S"
>    THEN
>       OUTPUT Num1 - Num2
> ELSEIF Choice = "M"
>    THEN
>       OUTPUT Num1 * Num2
> ELSEIF Choice = "D"
>    THEN
>       OUTPUT Num1 / Num2
>    ELSE
>       OUTPUT "Incorrect entry"
> ENDIF
> ```
>
> The pseudocode shows the CASE statement and ELSEIF approach follow a similar format. Compare these with the nested IF approach to see how CASE or ELSEIF improves readability. The declaration and input code has been omitted as it is identical to the ELSEIF statement.
>
> ```
> IF Choice = "A"
>   THEN
>     OUTPUT Num1 + Num2
>   ELSE
>     IF Choice = "S"
>       THEN
> ```

> **CONTINUED**

```
            OUTPUT Num1 - Num2
      ELSE
        IF Choice = "M"
          THEN
            OUTPUT Num1 * Num2
          ELSE
            IF Choice = "D"
              THEN
                OUTPUT Num1 / Num2
              ELSE
                OUTPUT "Incorrect entry"
            ENDIF
        ENDIF
    ENDIF
ENDIF
```

The following code is a console application implementation of the CASE statement:

```vb
Module Module1
    Sub Main()
        Dim Num1 As Decimal = 0
        Dim Num2 As Decimal = 0
        Dim Choice As String = ""

        Console.WriteLine("Enter first number")
        Num1 = Console.ReadLine()
        Console.WriteLine("Enter operator required")
        Console.WriteLine("A = Add, S = Subtract, M = Multiply, D = Divide")
        Choice = Console.ReadLine()
        Console.WriteLine("Enter second number")
        Num2 = Console.ReadLine()

        Select Case Choice
            Case "A"
                Console.WriteLine(Num1 + Num2)
            Case "S"
                Console.WriteLine(Num1 - Num2)
            Case "M"
                Console.WriteLine(Num1 * Num2)
            Case "D"
                Console.WriteLine(Num1 / Num2)
            Case Else
                Console.WriteLine("Incorrect operator")
        End Select
        Console.ReadKey()
    End Sub
End Module
```

4 Selection

CONTINUED

This approach requires the user to input the first letter of the operator. Alternative approaches would work equally efficiently. For example, the user could be required to input 1 for Add, 2 for Subtract, etc. The CASE statement would then use an Integer variable instead of a string.

Note: Windows Forms application is optional content. It is not covered in the syllabuses.

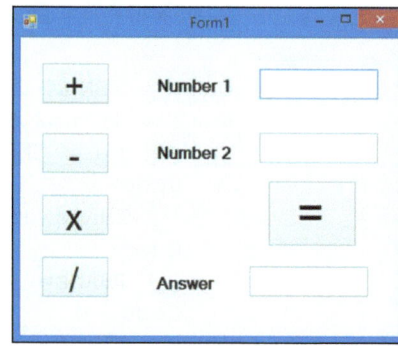

For a Windows Forms application, a GUI could be used to input data. Instead of typing in the required operation name, the user could select from a choice of buttons (see Figure 4.1). The design for the interface will need to contain the input and output textboxes, the four operation buttons and an 'equals' button:

The symbol on each button has been achieved by inputting the mathematical symbol as the button text. The font of each button has then been increased using the properties window.

Figure 4.1: A GUI for the calculator program

It is important that the Choice variable is declared as a global variable (declared outside of any subroutine) to give scope to all the button routines.

Because the operator choice is made by selecting different buttons, unlike the console application version where the operator is typed in, it is impossible to select an incorrect operator. It is, however, possible to select the equals button before an operator choice has been made. If the user has failed to select an operator, the string variable Choice will still hold an empty string. Should this happen, the CASE ELSE condition will execute and output the message 'Select operator'.

One advantage of using the Windows Forms application over the console application is that the Windows Forms application form allows for multiple calculations to be completed – the form will not close until the application is exited. The console application will close at the end of each calculation. It is possible to make the console application remain open and offer the user multiple calculations by using a WHILE loop (see Chapter 5).

```
Public Class Form1
    Dim Choice As String = ""

    Private Sub BTNAdd_Click(sender As Object, e As EventArgs) Handles BTNAdd.Click
        Choice = "A"
    End Sub
    Private Sub BTNSubtract_Click(sender As Object, e As EventArgs) Handles BTNSubtract.Click
        Choice = "S"
    End Sub
    Private Sub BTNMultiply_Click(sender As Object, e As EventArgs) Handles BTNMultiply.Click
        Choice = "M"
    End Sub
    Private Sub BTNDivide_Click(sender As Object, e As EventArgs) Handles BTNDivide.Click
        Choice = "D"
    End Sub
```

CONTINUED

```vb
    Private Sub BTNEquals_Click(sender As Object, e As EventArgs) Handles
      BTNEquals.Click
        Dim Num1 As Decimal = 0
        Dim Num2 As Decimal = 0

        Num1 = TBNum1.Text
        Num2 = TBNum2.Text
        Select Case Choice
            Case "A"
                TBAnswer.Text = Num1 + Num2
            Case "S"
                TBAnswer.Text = Num1 - Num2
            Case "M"
                TBAnswer.Text = Num1 * Num2
            Case "D"
                TBAnswer.Text = Num1 / Num2
            Case Else
                ("Select Operator")
        End Select
    End Sub
End Class
```

CHALLENGE TASK 4.2

Note: Windows Forms application is optional content. It is not covered in the syllabuses.

This task can only be implemented as a Windows Forms application.

The calculator you have made requires you to input two numbers into two different input textboxes. Most calculators only have one display box for both the input and output. They follow this process:

1 Input first number in display box.

2 Select operator. This clears the display box and stores the first number.

3 Input second number in display box.

4 Select 'equals'. This performs the intended operation and displays the result.

5 Select 'clear', which clears stored values and clears the display.

Produce a flowchart or pseudocode to create a program for a calculator that has only one textbox and performs the process described above.

Using CASE with scaled values

CASE statements work in a similar way to ELSEIF statements. The criteria can be a mathematical calculation. The statement will check each case in turn to identify if the criteria is TRUE. If the criteria is FALSE, the statement will continue to check the remaining cases. When a TRUE criteria is met, the appropriate code will be executed, and the statement will be exited. CASE statements can therefore be used with scaled values in the same way that ELSEIF statements are used.

A system is required to calculate and output an examination grade using the grade boundaries shown in Table 4.4.

Grade	Boundary
Merit	Score above 25
Pass	Score between 15 and 24
Fail	Score below 15

Table 4.4: Grade boundaries for CASE statement

```
Select Case Score
    Case > 25
        Console.WriteLine("Merit")
    Case >= 15
        Console.WriteLine("Pass")
    Case Else
        Console.WriteLine("Fail")
End Select
```

This CASE statement approach to completing this task makes used of the same logic and criteria as the ELSEIF statement approach explained in Section 4.6.

> **TIP**
>
> It is possible to code a given scenario by making use of any of the selection methods. When deciding on the best method, consider the advantages of each statement:
> - CASE…OTHERWISE…END CASE offers an easy-to-read selection where multiple paths are determined by a single variable or input. Only one path is followed.
> - IF…THEN…ELSE…END IF offers a simple solution where only two paths exist. It can have complex criteria consisting of several variables or inputs.
> - IF…THEN…ELSEIF…ELSE…END IF allows multiple paths to be considered. It can have complex criteria consisting of several variables or inputs.
> - Nested IF allows multiple paths to be considered. Subsequent decisions can be made depending on earlier decisions. It can have complex criteria consisting of several variables or inputs.
>
> All the options offer a default path that is executed when all the criteria are false.

> **PRACTICE TASKS 4.6–4.7**
>
> | Grade Achieved | Grade Boundary |
> |---|---|
> | A | More than 100 marks |
> | B | Between 80 and 99 marks |
> | C | Between 60 and 79 marks |
> | D | Between 40 and 59 marks |
> | U | Less than 40 marks |
>
> **4.6a** Design the flowchart or pseudocode of a system that calculates the grade a student has achieved in an examination. The maximum number of marks in the examination is 120. The system will take as input the number of marks achieved by the student and output the grade achieved using the grade boundaries shown.
>
> **b** Using a CASE statement, implement and test your algorithm in a Visual Basic console application or Windows Forms application.
>
> **c** Reproduce your algorithm using a series of nested IF statements.
>
> **4.7a** Design the flowchart or pseudocode of a suitable algorithm for a system that will output the area of the three shapes detailed in the table. The user will select a shape and then they will be prompted to input the data required to calculate the area of the shape they have selected. The system will output the area of the selected shape.
>
> | Shape | Data required | How to calculate the area |
> |---|---|---|
> | Rectangle | Length
Width | Area = Length * Width |
> | Triangle | Length of base
Perpendicular height | Area = Base Length * perpendicular height / 2 |
> | Circle | Radius | Pi * Radius^2 (Pi = 3.14159) |
>
> **b** Using a CASE statement implement and test your algorithm in a Visual Basic console application or Windows Forms application.
>
> **c** Reproduce your algorithm using an ELSEIF statement.

4.9 Creating complex conditions

Often a single criteria is not sufficient to define the required condition. For example, a fire alarm system may be required to activate if it detects either the presence of smoke or a high temperature. The condition in this case requires two criteria, either of which being true would cause activation of the alarm.

Visual Basic, in common with many other languages, uses the Boolean operators (shown in Table 4.5) to create complex conditions.

Operator	Description	Example
AND	All connected operators must be True for the condition to be met.	`IF StudentUser = True  AND IDNumber > 600` The condition will only be True where the user is a student with an ID number higher than 600. Any other type of user with an ID number > 600 will not meet the criteria because it will fail the `StudentUser = True` element of the condition.
OR	Only one of the connected operators needs to be True for the condition to be met.	`IF SmokeDetected = True  OR Temperature > 70 °C` The condition will be True if either smoke is detected or the temperature is above 70 °C. It will also be True if both elements are met.

NOT	Used where it is easier to define the logical criteria in a negative way.
	
	Can also be written as <>
		`IF NOT Input_Number = 6`
		`IF Input_Number <> 6`
		
		The condition will be True for any number input with the exception of the number 6.

Table 4.5: Boolean operators

> **KEY WORDS**
>
> **AND Boolean operator:** used to create a condition which is made up from more than one criteria. The condition will only be TRUE if **all** of the individual criteria are TRUE.
>
> **OR Boolean operator:** used to create a condition which is made up from more than one criteria. The condition will be TRUE if **any** of the individual criteria are TRUE.
>
> **NOT Boolean operator:** used to create a condition which will be TRUE when the criteria are FALSE.

> **TIP**
>
> The Boolean operators AND and OR act in a similar way to the AND and OR logic gates. The AND operator means both criteria must be true for the connected criteria to be true. Similar to the way that an AND gate requires both inputs to be on for the output to be on. The OR operator mirrors the function of an OR gate because either/or both criteria being true will result in the connect criteria being true.

4.10 Using Boolean operators to connect criteria

One way to test multiple criteria is to use IF...ELSEIF or nested IF statements. It is common practice to use Boolean operators to create complex conditions that replace the complex IF statements. The complex condition will test each of the connected criteria in one statement as an alternative to testing each criteria in a series of statements.

Connected criteria cannot be used with CASE statements as the CASE can only test a single criteria.

Showing connected criteria in a flowchart is achieved by writing the connect criteria test in the single decision symbol. Flowchart 4.6 shows the design for a system that requires an input number to be a positive even number.

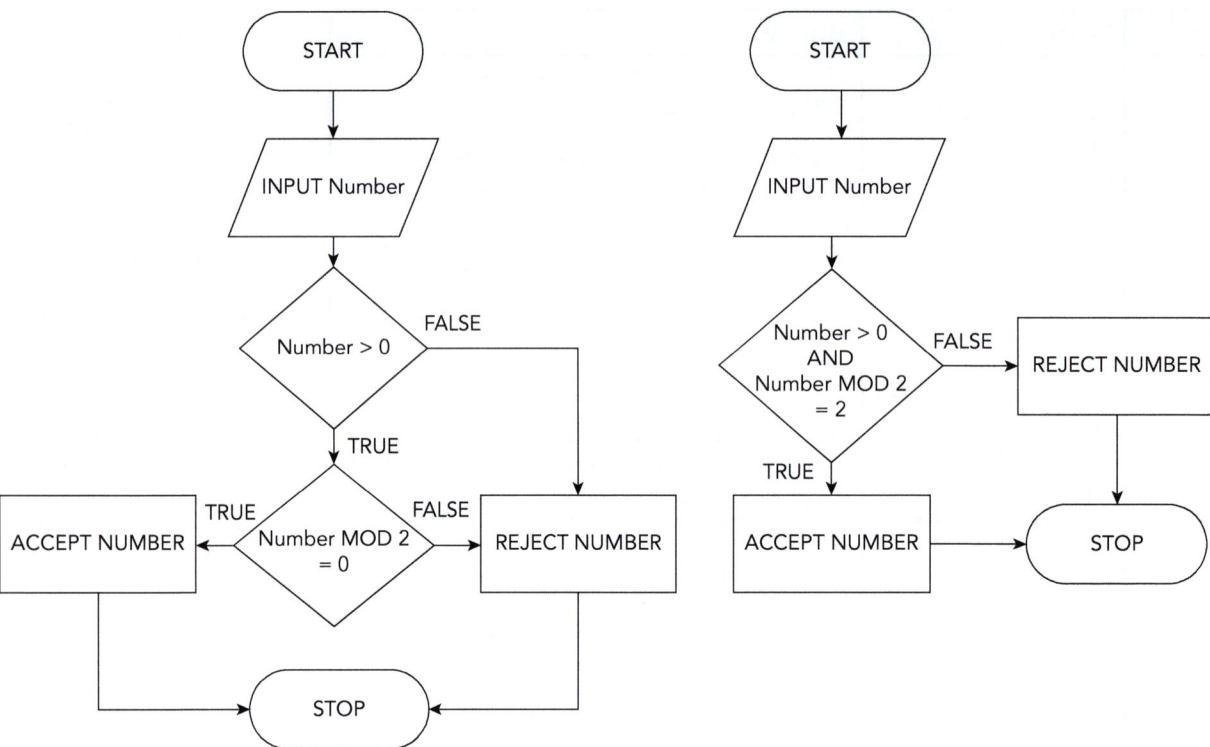

Flowchart 4.6: Different flowchart approaches to the positive even number task

The approach of connecting the criteria is also used in the coded implementation of this algorithm. Two approaches are shown in Code snippet 4.6.

```
Dim Number As Integer
Number = Console.ReadLine()

If Number > 0 Then
    If Number Mod 2 = 0 Then
        'accept number'
    Else
        'reject number'
    End If
Else
    'reject number
End If
```

```
Dim Number As Integer
Number = Console.ReadLine()

If Number > 0 And Number Mod 2 = 0 Then
    'accept number
Else
    'reject number
End If
```

Code snippet 4.6: Different coded approaches to the positive even number task

Table 4.6 shows a series of examples of the use of connected criteria.

Task	Individual Criteria	Connected Criteria

A 10% discount is available to young people or to old people.
 	
Children aged under 16 or people aged over 65.
	`Dim Cost As Decimal = 0`
	`Dim Age As Integer = 0`
	`Age = Console.ReadLine()`
	
	`If Age < 16 Then`
	`    Cost = Cost * 0.9`
	`ElseIf Age > 65 Then`
	`    Cost = Cost * 0.9`
	`End If`
		`Dim Cost As Decimal = 0`
		`Dim Age As Integer = 0`
		`Age = Console.ReadLine()`
		
		`If Age < 16 Or Age > 65 Then`
		`    Cost = Cost * 0.9`
		`End If`

A number input must be above 1 but cannot be number 13.	`Dim Number As Integer = 0`
	`Number = Console.ReadLine()`
	
	`If Number > 1 Then`
	`    If Number <> 13 Then`
	`        'accept number'`
	`    Else`
	`        'reject number'`
	`    End If`
	`Else`
	`    'reject number'`
	`End If`
		`Dim Number As Integer = 0`
		`Number = Console.ReadLine()`
		
		`If Number > 1 And`
		`        Number <> 13 Then`
		`    'accept number'`
		`Else`
		`    'reject number'`
		`End If`

The command `Not Number = 13` could have been used in place of `Number <> 13`	

A number input must be between 10 and 20
 	
This task can be achieved in two different ways:
 	
Check the Number is within the acceptable range before accepting the number.
 	
`If Number <= 20 And`
`        Number >= 10`
`Then`
	`Dim Number As Integer = 0`
	`Number = Console.ReadLine()`
	
	`If Number <= 20 Then`
	`    If Number >= 10 Then`
	`        'accept number'`
	`    Else`
	`        'reject number'`
	`    End If`
	`Else`
	`    'reject number'`
	`End If`
		`Dim Number As Integer = 0`
		`Number = Console.ReadLine()`
		
		`If Number <= 20 And`
		`        Number >= 10 Then`
		`    'accept number'`
		`Else`
		`    'reject number'`
		`End If`

(continued)

Task	Individual Criteria	Connected Criteria

Check the number is outside the acceptable range before rejecting the number.
 	
`If Number < 10 Or`
`        Number > 20`
`Then`
	`Dim Number As Integer = 0`
	`Number = Console.ReadLine()`
	
	`If Number < 10 Then`
	`    'reject number'`
	`Else`
	`    If Number > 20 Then`
	`        'reject number'`
	`    Else`
	`        'accept number'`
	`    End If`
	`End If`
		`Dim Number As Integer = 0`
		`Number = Console.ReadLine()`
		
		`If Number < 10 Or`
		`        Number > 20 Then`
		`    'reject number'`
		`Else`
		`    'accept number'`
		`End If`

The criteria in both approaches would have been equally effective if the order was reversed:
 	
`If Number >= 10 And Number <= 20 Then`
`If Number > 20 Or Number < 10 Then`		

Table 4.6: Examples of connected criteria

> **TIP**
>
> The English language often uses the words AND and OR differently to the way they are applied as logical operators. If the first example in Table 4.6 had been phrased 'Both children and the elderly will get a discount', you would have understood the meaning of the phrase. But using the connector from the sentence in the code `If Age < 16 And Age > 65 Then` would be nonsense as it is impossible to be both ages.
>
> Be careful to examine the logic of a situation and not be influenced by the connection words used in the description of the situation.

4.11 Connecting more than one value

Table 4.6 shows examples of connecting criteria based on one value. It is also possible to connect multiple values in one complex criteria.

> **DEMO TASK 4.3**
>
> *A system is required to calculate the delivery cost of parcels. All parcels below 5 kg in weight have a fixed charge for delivery. For local deliveries, this charge is $20; for international deliveries, the charge raises to $40.*
>
> *The maximum weight limit for international deliveries is 5 kg, parcels in excess of 5 kg are rejected. However, for local deliveries, extra weight is accepted and is charged at $1 for every kilogram or part kilogram above the 5 kg limit.*
>
> *Flowchart 4.7 shows the logic of the system. It makes use of connected criteria based on the weight and of the parcel and the delivered type.*

4 Selection

CONTINUED

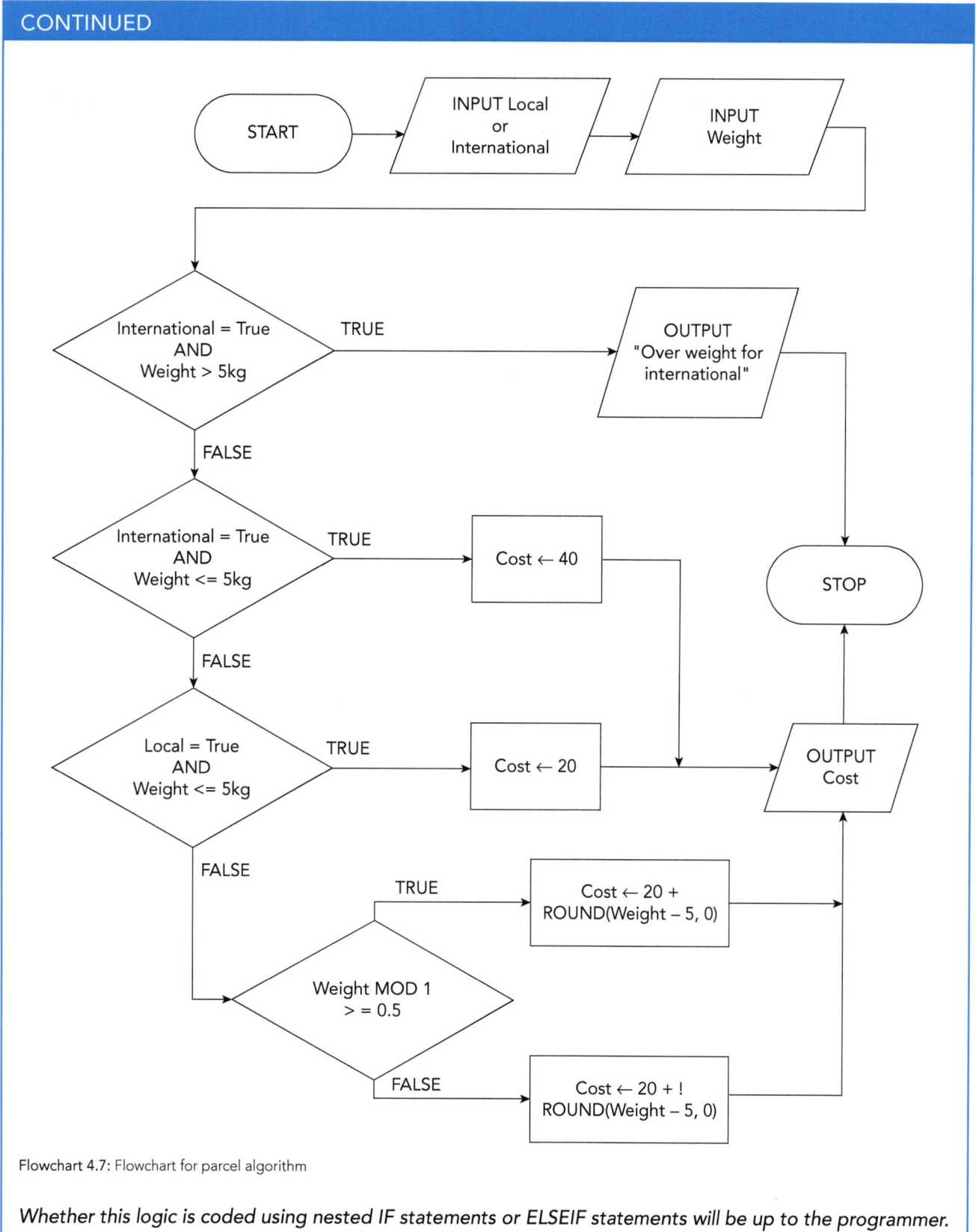

Flowchart 4.7: Flowchart for parcel algorithm

Whether this logic is coded using nested IF statements or ELSEIF statements will be up to the programmer.

CONTINUED

Solution

A console application nested IF solution to this task could be implemented as shown in Code snippet 4.7.

```vb
Module Module1
    Sub Main()
        Dim DeliveryType As String = ""
        Dim Weight As Decimal = 0
        Dim Cost As Decimal = 0
        Const IntCost As Integer = 40
        Const LocCost As Integer = 20

        Console.WriteLine("Select delivery type")
        Console.WriteLine("L = Local,  I = International")
        DeliveryType = Console.ReadLine()
        Console.WriteLine("Enter weight of parcel")
        Weight = Console.ReadLine()

        If DeliveryType = "I" Then
            If Weight > 5 Then
                Console.WriteLine("REJECTED - Overweight for international
                    delivery")
            Else
                Console.WriteLine(IntCost)
            End If
        Else
            If Weight <= 5 Then
                Console.WriteLine(LocCost)
            Else
                Cost = LocCost + Math.Ceiling(Weight - 5)
                Console.WriteLine(Cost)
            End If
        End If
        Console.ReadKey()
    End Sub
End Module
```

Code snippet 4.7: nested IF implementation of the parcel algorithm

An alternative solution would replace the nested IF with an ELSEIF statement using connected criteria. The code for declaring variables and constants as well as obtaining of user input would remain unaltered, as shown in Code snippet 4.8.

```vb
If DeliveryType = "I" And Weight > 5 Then
    Console.WriteLine("REJECTED - Overweight for international delivery")
ElseIf DeliveryType = "I" And Weight <= 5 Then
    Console.WriteLine(IntCost)
ElseIf DeliveryType = "L" And Weight <= 5 Then
    Console.WriteLine(LocCost)
```

CONTINUED

```
Else
    Cost = LocCost + Math.Ceiling(Weight - 5)
    Console.WriteLine(Cost)
End If
```
Code snippet 4.8: ELSEIF with connect criteria implementation of the parcel algorithm

TIP

Demo Task 4.3 requires the program to round up to the next whole kilogram. The pseudocode and Visual Basic approach to rounding is covered in Chapter 3.

SUMMARY

Selection provides methods that programmers can use to allow the algorithm to follow different paths through the code depending on the data being used at the time.
The flowchart symbol for selection is a decision diamond. The selection criteria are included in the symbol. The exit paths should be indicated as TRUE and FALSE.
Multiple decisions are shown as a series of connected decisions.
Logical operators are used to provide a range of comparative options.
nested IF statements provide the ability for additional conditions to be checked once earlier conditions have determined a path.
ELSEIF statements are an alternative to nested IF statements. Each condition is tested in turn. When a condition is TRUE, the code related to that condition is executed and the statement is ended.
CASE statements provide a simple method of providing multiple paths based on a single variable or user input.
The ELSE statement (used with IF) and the OTHERWISE statement (used with CASE) provide a default path should none of the conditions be met.
Boolean operators can be used with IF statements to provide the ability to use complex conditions based on multiple variables or user inputs: • AND – will connect multiple criteria to make a complex condition. The condition will be TRUE only when all of the individual criteria are TRUE. • OR – will connect multiple criteria to make a complex condition. The condition will be TRUE if any of the individual criteria are TRUE. • NOT – used to create a condition which will be TRUE when the criteria are FALSE.

END-OF-CHAPTER TASKS

1 a Using either flowchart or pseudocode, design an algorithm for a system that takes as input three different integer values. The system will output the highest value input. If any of the integer values match, the system will output an error message.

   b Implement and test your algorithm using a nested IF solution.

   c Implement and test your algorithm using an ELSEIF solution.

2 Implement an algorithm to the earlier leap year task which uses a single IF – ELSE – END IF statement with connected criteria. The system will take as an input the year of the person's birth and outputs either 'Leap Year' or 'Normal Year'.

   The process for identifying if a year is a leap year is:

   Step 1 – If the year is evenly divisible by 4 then go to step 2, otherwise go to step 5

   Step 2 – If the year is evenly divisible by 100 then go to step 3, otherwise go to step 4

   Step 3 – If the year is evenly divisible by 400 then go to step 4, otherwise go to step 5

   Step 4 – The year is a leap year and has 366 days

   Step 5 – The year is not a leap year and has 365 days

3 a Using either flowchart or pseudocode, design an algorithm for a system that is used to calculate the cost of a ticket for a UK airline. The airline is based at Gatwick and only flies to two other locations in the UK. The airline offers both economy and business class tickets. Details of the flight costs are shown in the table.

Location / Extras	Economy		Business Class	
Edinburgh	One Way	£32	One Way	£87
	Return	£52	Return	£178
Cardiff	One Way	£36	One Way	£72
	Return	£50	Return	£158
In-flight meal	One Way	£12	Included in the price	
	Return	£18		

The system will obtain all the required information from the passenger before outputting the total cost of the ticket selected.

   b Implement and test your algorithm using either a console application or Windows Forms application.

4 Selection

> CONTINUED
>
> **4** A system is required that will calculate the cost of a guest staying at a small hotel. The system will take as inputs:
> - the number of nights the guest intends to stay
> - if the guest wants breakfast included
> - whether the guest wants a deluxe or a standard room.
>
> The system will output the total cost of the stay using these cost factors.
>
> | Details | Costs |
> |---|---|
> | Standard room | $40 per night for stays of 1 to 3 nights |
> | | $35 per night for stays of longer than 3 nights |
> | Deluxe room | $60 per night for stays of 1 to 3 nights |
> | | $45 per night for stays of longer than 3 nights |
> | Breakfast | $10 for each night of stay |
>
> **a** Design a suitable algorithm in pseudocode or a flowchart.
>
> **b** Implement and test your algorithm in a Visual Basic console application or Windows Forms application.

Chapter 5
Iteration

IN THIS CHAPTER YOU WILL:

- learn about the need for iteration
- know how to design and represent iteration using flowcharts and pseudocode
- be able to write code that will repeat instructions a predetermined number of times
- be able to write code that will repeat instructions based on user input
- know how to use counters with repeated code
- understand the advantages and disadvantages of FOR, WHILE and REPEAT…UNTIL loops
- understand how to use nested iteration.

5 Iteration

Introduction

There are many tasks that require us to repeat a series of actions several times. For example, when you learnt how to multiply in mathematics, your teacher probably will have given you lots of different numbers to multiply together so you could practise the skill. Although the numbers changed, you will have repeated the same series of steps for each of the numbers.

Iteration is the programming equivalent of repetition. It is not unusual for a program to deal with a series of inputs, each input requiring the same process to be applied. Iteration allows a section of code to be repeated multiple times with the values changing at each iteration. This is more efficient than writing out the code multiple times.

> **KEY WORD**
>
> **iteration:** code repeats a certain sequence a number of times depending on certain conditions.

5.1 Types of iteration

Three basic forms of **iteration** (see Table 5.1) exist in the majority of programming languages. They are known as 'loops', because they cause the program to repeatedly 'loop through' the same lines of code.

Loop type	Description	When it should be used
FOR loop	Repeats a section of code a predetermined number of times. These are known as count-controlled loops because a counter is used to control the number of iterations	The number of iterations is known or can be calculated. The programmer can set the code to loop the correct number of times.
WHILE loop	Repeats a section of code while the control condition is true. These are also known as pre-condition loops because the condition is checked before the loop runs.	The number of iterations is not known and it may be possible that the code will never be required to run. The condition is checked before the code is executed. If the condition is false, the code in the loop will not be executed.
REPEAT... UNTIL loop	Repeats a section of code until the control condition is true. These are also known as post-condition loops because the condition is checked after the loop has run once.	The number of iterations is not known but the code in the loop must be run at least once. The condition is checked after the code has been executed, so the code will run at least once.

Table 5.1: Three basic forms of iteration

Often it is possible to use any of the three types when producing an algorithm. However, each type offers the programmer certain advantages. Selecting the most appropriate type of loop can help to make your code more efficient.

> **KEY WORD**
>
> **FOR loop:** a type of iteration that will repeat a section of code a known number of times. Also known as a count-controlled loop.

5.2 FOR loops (count-controlled loops)

A **FOR loop** can only be used where the number of iterations is known. This can happen in a situation either where the number of iterations is fixed ('hard coded' when it is programed) or when a variable is given a value by a user and that value is used to determine the number of iterations. Table 5.2 gives some examples.

Type	Example	Reason for use
Fixed number of iterations.	Loop code to check the figures entered for every month in the year.	There are always 12 months in any year. The loop could be 'hard-coded' to loop 12 times.
Number of iterations determined by a variable.	Loop code to check the figures entered for every day in a month.	Months have different number of days, ranging from 28 to 31. As the program could be run for any month, the number of iterations cannot be 'hard-coded'. The system would need the user to input the number of days in the month they wanted to check.

Table 5.2: Reasons for using a FOR loop

FOR loops are also known as 'count-controlled' loops because the number of iterations is controlled by a **loop counter**.

When writing a FOR loop in Visual Basic, you need to follow this format:

```
For Counter = 1 To 10
    'Code to be executed 10 times
Next
```

Each individual element of the loop performs an important role in achieving the iteration as shown in Table 5.3.

> **KEY WORD**
>
> **loop counter:** a variable that is used within a FOR loop to keep a record of the amount of times the loop has been repeated. The loop counter normally increases by 1 each time the loop is executed.

Element	Description
`For`	The start of the loop.
`Counter = 1 To 10`	`counter` is a variable that records the number of iterations that have been run. This is usually incremented (increased) by 1 every iteration. In Visual Basic, you do not need to declare the counter variable separately – it is automatically declared as part of the FOR loop. The value of the counter variable can be used within the loop to perform incremental calculations.
`Next`	The end of the iteration section. The value of the `counter` variable is incremented, and the flow of the program goes back to the start of the loop. The loop criteria will check to see if the counter value is within the condition (10 in this example). If the counter has exceeded the end value, the loop will direct the program to the line of code following the end of the loop (the `Next`); if not, it will rerun the loop. It is a common misconception that once the maximum number of iterations has been reached (in this case, 10), the `Next` will exit the loop. This is not true. The FOR loop in the example is written to execute 10 times. Although the loop counter may have reached 10, `Next` will still increment to counter to 11 before the program returns to the criteria check of `For counter = 1 To 10`. This will determine that the value of the loop counter is outside the criteria and will then exit the loop.

Table 5.3: Explanation of the FOR loop code

5.3 Using the loop counter

The loop counter maintains a count of the number of iterations the FOR loop has completed. One of the advantages of a FOR loop is that the value held in the loop counter can be used directly in any code within the loop.

Consider a program that is required to output the numbers 1 to 50.

The steps needed to complete this task are:

1. Repeatedly increment the value in a variable by 1.
2. Output the value in variable each time it is incremented.
   - Repeat steps 1 and 2 for a maximum of 50 times.

You might consider using the following pseudocode approach:

```
DECLARE Number : INTEGER ← 0
FOR Counter ← 1 TO 50
    Number ← Number + 1
    OUTPUT Number
NEXT Counter
```

This code would work, but the variable `Number` is not necessary. The loop counter variable `Counter` is already being incremented each time the loop iterates to keep a count of how many times the loop has run. The variable `Counter` could be used directly in the code.

```
FOR Counter ← 1 TO 50
    OUTPUT Counter
NEXT Counter
```

Console application solution

```
Module Module1
    Sub Main()
        For Counter = 1 To 50
            Console.WriteLine(Counter)
        Next
        Console.ReadKey()
    End Sub
End Module
```

It is common for programs to use the loop counter to achieve tasks that require incremental processes to take place.

DEMO TASK 5.1

A system is required to output the multiples of a given number up to a maximum of 10 multiples. For example, the multiples of 6 are 6, 12, 18, 24, 30, 36, 42, 48, 54 and 60.

Solution

Flowchart 5.1 shows the design of the algorithm. The pseudocode is also shown. Although the counter is automatically declared in Visual Basic, this is not the case with all languages, so it is normal to include the declaration in the design.

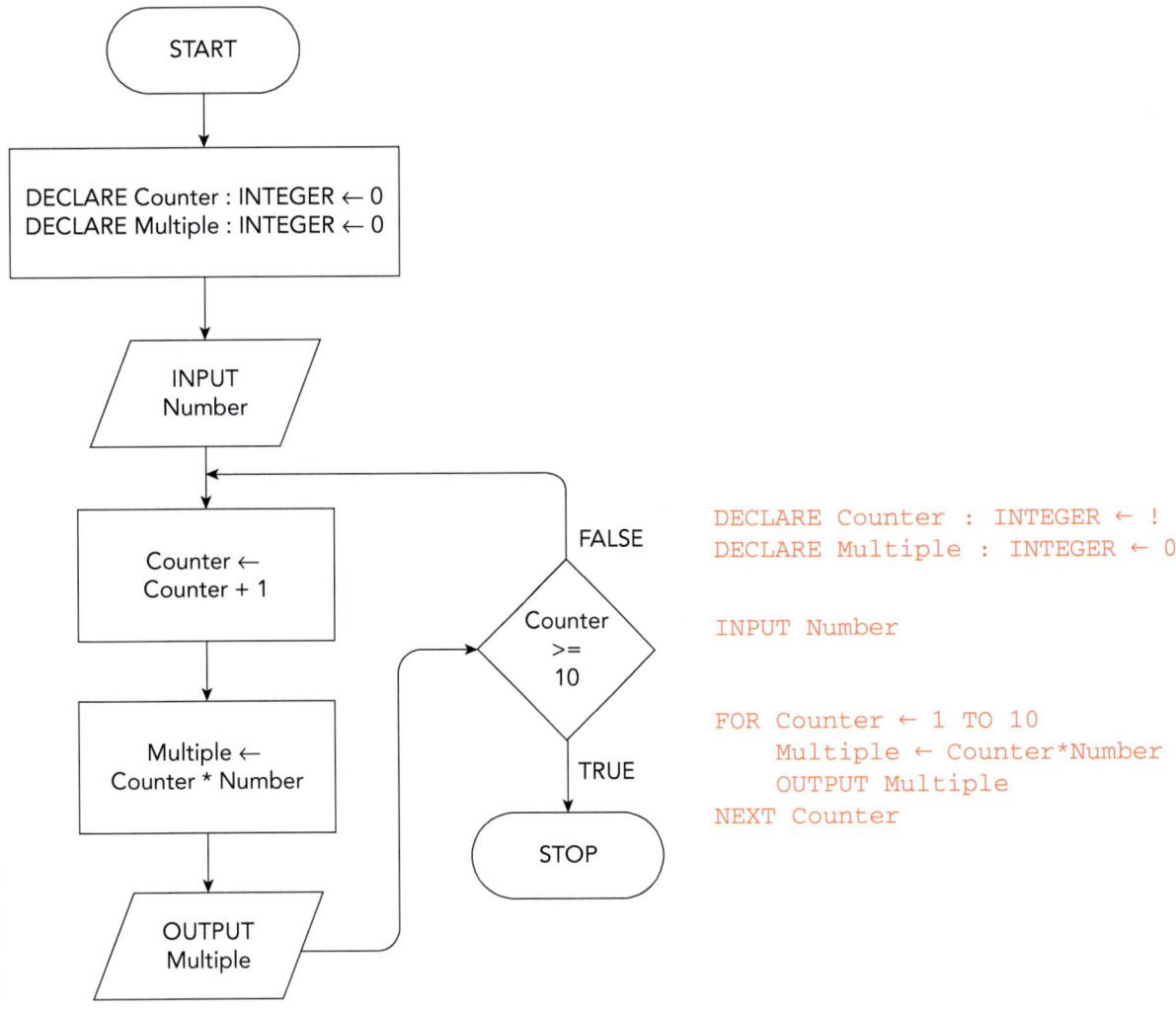

Flowchart 5.1: Flowchart for outputting multiples

The code for the console application for this task could be similar to the following:

```
Module Module1
    Sub Main()
        Dim Multiply As Integer

        Console.WriteLine("Input number to Multiply")
```

CONTINUED

```
        Multiply = Console.ReadLine()


        For Counter = 1 To 10
            Console.WriteLine(Multiply * Counter)
        Next
        Console.ReadKey()


    End Sub
End Module
```

Note: Windows Forms application is optional content. It is not covered in the syllabuses.

The Windows Forms application in Figure 5.1 makes use of a list box that will hold multiple string entries as a list of items.

Items are added to the listbox by using the code:

```
ListboxName.Items.Add("item To add")
```

The item to add can be a textual value enclosed by double quotation marks, for example "textual value". Alternatively, the name of a variable can be provided, in which case the value in the variable is used.

The code to achieve this task would be:

```
Public Class Form1
'Declare and initialise the variable
to hold the input number
Dim Multiply As Integer


Private Sub BTNMultiply _ Click(sender
As Object, e As EventArgs)
Handles BTNMultiply.Click
    'Obtain the input value from the textbox and place in the variable
    Multiply = TBInput.Text
    'Start the FOR loop to iterate 10 times
    For Counter = 1 To 10
    'Add the calculated multiplication to the list box
        ListBoxOutput.Items.Add(Multiply * Counter)
    Next
End Sub
End Class
```

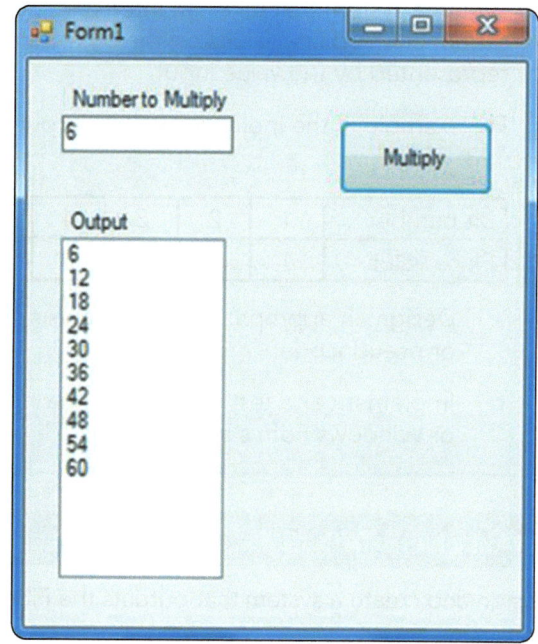

Figure 5.1: Windows Forms application of multiplier

PRACTICE TASKS 5.1–5.2

5.1 Create a system that will take an integer value as an input and output all the square numbers up to that value (a square number is a number multiplied by itself).

For example, if the input was 5, the system would output 1, 4, 9, 16, and 25.

Number	1	2	3	4	5
Square	1	4	9	16	25

- **a** Design an appropriate algorithm using either flowchart or pseudocode.
- **b** Implement and test your solution using either a console application or Windows Forms application.

5.2 Create a system that takes an integer value as an input. The system will output all the place values on the binary line up to the number of bits represented by the value input.

For example, if the input was 9, the output would be 1, 2, 4, 8, 16, 32, 64, 128, 256.

Bit number	1	2	3	4	5	6	7	8	9
Place value	1	2	4	8	16	32	64	128	256

- **a** Design an appropriate algorithm using either flowchart or pseudocode.
- **b** Implement and test your solution using either a console application or Windows Forms application.

> **TIP**
>
> In Practice Task 5.2, each value is double the previous value (increased by the power of 2).
>
> Your solution should display the values as a vertical list. The table shown here is included only to help explain the task.

CHALLENGE TASK 5.1

Design and create a system that outputs the Fibonacci sequence.

The Fibonacci sequence is a series of numbers where each number is the sum of the two previous numbers. The sequence starts with the values 0 and 1. The sequence for the first eight numbers is 0, 1, 1, 2, 3, 5, 8, 13.

The system will take an integer value that represents the nth value in the sequence. The system will output the value of the Fibonacci sequence up to the nth value.

For example, If the input value (the nth value) was 10, the system, would output the first ten numbers in the sequence: 0, 1, 1, 2, 3, 5, 8, 13, 21, 34.

5.4 Combining iteration and selection

You will find situations where a system will be required to loop code for several different input values. However, instead of repeating an identical process at each iteration, the system will need to perform a different action dependent on the result of a calculation or comparison. For example, for a system that is required to check a series of carbon dioxide (CO_2) readings from a water quality test and output any results that are higher than the acceptable CO_2 level, the following two programming concepts would be used:

- Iteration (a FOR loop) to check every reading.
- Selection (an IF statement) to identify and output any reading higher than the acceptable value.

The pseudocode format to achieve this type of system would be:

```
DECLARE NumberOfReadings : INTEGER ← 0
DECLARE AcceptableLevel : REAL ← 0
DECLARE Reading : REAL ← 0

OUTPUT "Input Number of Readings"
INPUT NumberOfReadings
OUTPUT "Input acceptable level"
INPUT AcceptableLevel
FOR Counter ← 1 to NumberOfReadings
   OUTPUT "Enter Reading"
   INPUT Reading
   IF Reading > AcceptableLevel
      THEN
         OUTPUT Reading
   ENDIF
NEXT Counter
```

DEMO TASK 5.2

A prime number can only be divided equally by 1 or itself. If a number can divide equally into another number, the modulus of that operation will be zero. Therefore, a prime number can also be defined as a number that when divided by all the positive integers between 1 and itself, will not result in a modulus of zero (e.g. 7 can only be divided by 1 or by 7; nothing else will divide into 7). We will use this rule to produce an algorithm capable of determining if a number is prime.

CONTINUED

Solution

Flowchart 5.2 shows the design of the algorithm. The pseudocode is also shown.

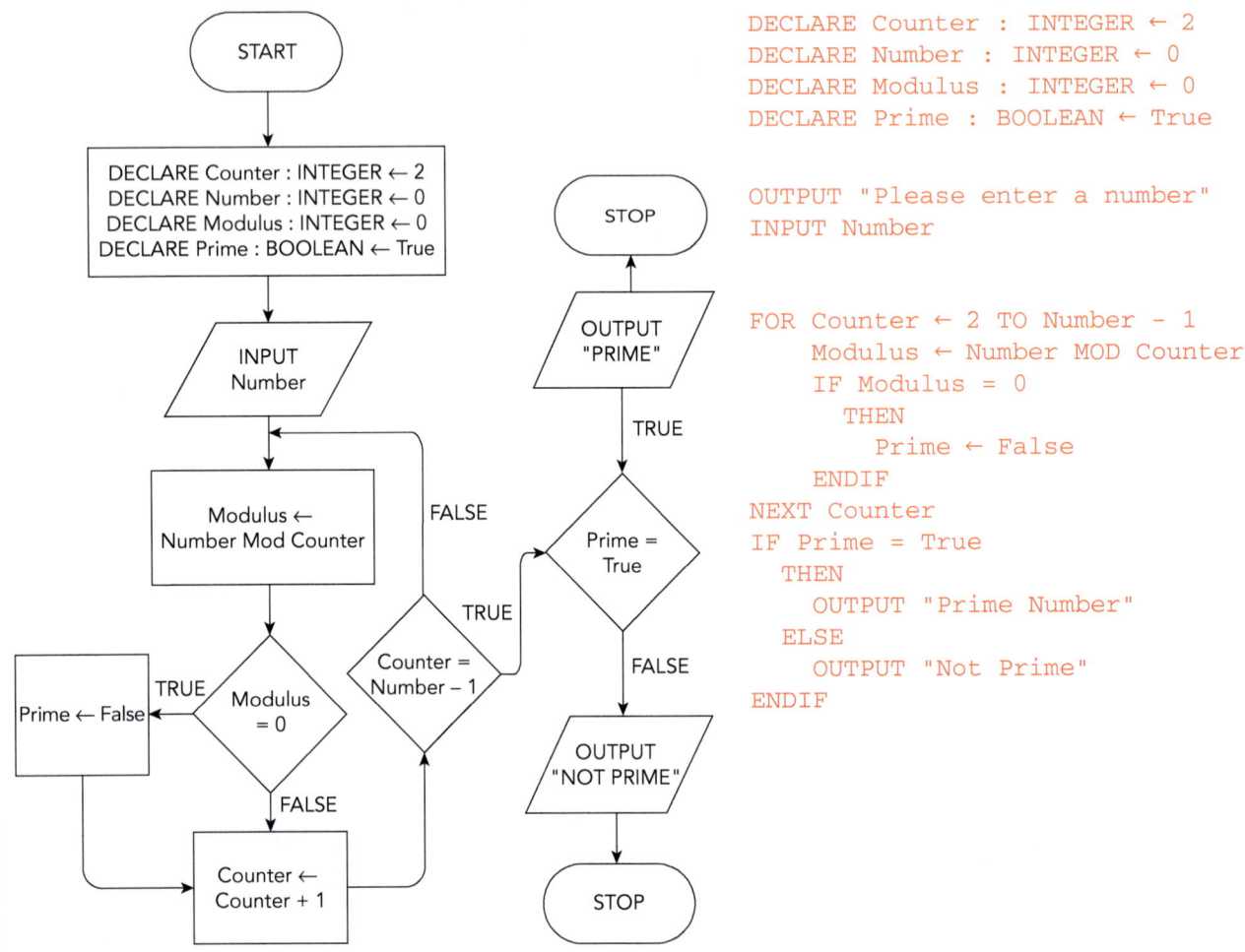

```
DECLARE Counter : INTEGER ← 2
DECLARE Number : INTEGER ← 0
DECLARE Modulus : INTEGER ← 0
DECLARE Prime : BOOLEAN ← True

OUTPUT "Please enter a number"
INPUT Number

FOR Counter ← 2 TO Number - 1
    Modulus ← Number MOD Counter
    IF Modulus = 0
       THEN
           Prime ← False
       ENDIF
NEXT Counter
IF Prime = True
   THEN
       OUTPUT "Prime Number"
   ELSE
       OUTPUT "Not Prime"
ENDIF
```

Flowchart 5.2: Flowchart for the prime number task

NOTE: In both designs, the iteration counter starts from 2 and ends at Number −1. This avoids the use of 1 and the number input in the loop as using these would result in a modulus of zero.

This is the console code that could be used to complete this task:

```
Module Module1
    Sub Main()
        'Declare and initialise variables
        Dim Number As Integer = 0
        Dim Modulus As Integer = 0
        Dim Prime As Boolean = True
        'Prime set to true as algorithm assumes number is prime until proven
        false
```

CONTINUED

```vb
        Console.WriteLine("Input Number to Test")
        Number = Console.ReadLine()


        'Start the loop from 2 to avoid using 1 in the
        'iteration. Range of loop is Number - 1 to avoid
        'using the input number in the iteration
        For Counter = 2 To Number - 1
            'Modulus for each positive number calculated
            Modulus = Number Mod Counter
            'IF statement will set Prime to False if the
            'modulus is zero
            If Modulus = 0 Then
                Prime = False
            End If
        'Next increments the counter and directs execution to FOR
        Next


        'Once FOR loop is complete Prime is used to
        'identify if input number was prime
        If Prime = True Then
            Console.WriteLine("This is a PRIME number")
        Else
            Console.WriteLine("This is NOT a PRIME number")
        End If
        Console.ReadKey()
    End Sub
End Module
```

> **Note:** Windows Forms application is optional content. It is not covered in the syllabuses.

For a Windows Forms application, the code would be similar, but the GUI could be designed as shown in Figure 5.2.

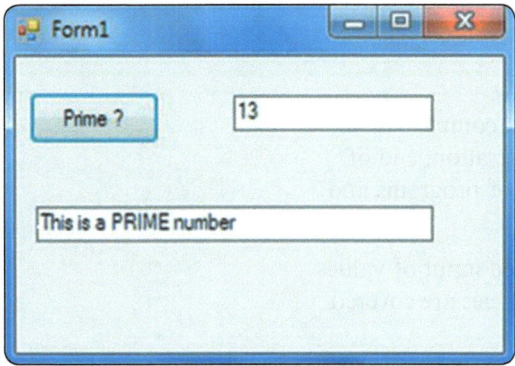

Figure 5.2: Windows Forms application of a prime number calculator

> **CAMBRIDGE IGCSE™ & O LEVEL COMPUTER SCIENCE: PROGRAMMING BOOK**

> **CONTINUED**
>
> The appropriate code is:
>
> ```vb
> Public Class Form1
>     Private Sub BTNPrime _ Click(sender As Object, e As EventArgs) Handles BTNPrime.Click
>         'Declare and initialise the modulus variables
>         'Declared locally so reset to zero if user runs code multiple times
>         Dim Modulus As Integer = 0
>         Dim ModCount As Integer = 0
>         Dim Number As Integer = 0
>
>
>         Number = TBInput.Text
>
>
>         For Counter = 2 To Number - 1
>             Modulus = Number Mod Counter
>
>
>             If Modulus = 0 Then
>                 Prime = False
>             End If
>         Next
>
>
>         If Prime = True Then
>             TBOutput.Text = "This is a PRIME number"
>         Else
>             TBOutput.Text = "This is a NOT a PRIME number"
>         End If
>     End Sub
> End Class
> ```
>
> **Discussion question:** Is limiting the iterations to Number - 1 the most efficient range limiter? What might be a better option?

5.5 Common iterative tasks

A combination of iteration and selection provides the basis for several common tasks. You will find examples of these tasks included in many of the demonstration, end of chapter and scenario tasks in this book. These are common elements of programs and you should become confident in their use.

The tasks involve working with multiple values, either through repeated input of values or through the use of arrays (data structures that can hold multiple values are covered in Chapter 10).

Table 5.4 shows the five common tasks.

Task	Description and pseudocode

Totalling	The process of calculating the total value of several values.
	
	Example task: A system is required to calculate the total value of 100 numbers input by a user.
	
	``
	DECLARE Total : REAL ← 0
	DECLARE Number : REAL ← 0
	FOR Counter ← 1 TO 100
	    OUTPUT "Enter a number"
	    INPUT Number
	    Total ← Total + Number
	NEXT Counter
	``
	
	The pseudocode will place the input of the number within a loop. Also, inside the loop each number is added to the current value in the variable `Total`. As each number is input the value in `Total` is increased by the value of the input.
	
	Some tasks will require the use of selection to only include certain values in the totalling process.
	
	Example Task: A system will allow the input of 50 numbers. It should output the total value of all numbers over 20.
	
	``
	DECLARE Total : INTEGER ← 0
	DECLARE Number : REAL ← 0
	FOR Counter ← 1 TO 50
	    OUTPUT "Enter a number"
	    INPUT Number
	    IF Number > 20
	      THEN
	         Total ← Total + Number
	      ENDIF
	NEXT Counter
	OUTPUT Total
	``
	
	Here the pseudocode makes use of an IF statement to ensure that only values over 20 are included in the totalling process. NOTE: the output is made once the loop has completed.

(continued)

Task	Description and pseudocode

Counting	The process of counting how many items have been input or meet a criteria.
	
	Example Task: A system is required that will allow the input of the marks in an examination for 600 students. The system should output how many students achieved less than 20 marks.
	
	``
	DECLARE MarksCount : INTEGER ← 0
	DECLARE Mark : INTEGER ← 0
	FOR Counter ← 1 TO 600
	    OUTPUT "Enter the mark for the student"
	    INPUT Mark
	    IF Mark < 20
	      THEN
	         MarksCount ← MarksCount + 1
	    ENDIF
	NEXT Counter
	OUTPUT MarksCounter
	``
	
	The pseudocode uses an IF statement to identify marks less than 20. If a mark less than 20 is entered the variable `MarksCounter` will be increment by 1. At the end of the process the variable will hold the number of times a mark less than 20 was entered. The code is effectively counting the number of marks less than 20 and this value is output once the loop has completed.

Finding maximum	The process of finding the largest (maximum) value of a series of values.
	
	Example Task: A system is required that will allow the input of the marks in an examination for 600 students. The system should output the highest mark achieved.
	
	``
	DECLARE Highest : INTEGER ← 0
	DECLARE Mark : INTEGER ← 0
	FOR Counter ← 1 TO 600
	    OUTPUT "Enter the mark for the student"
	    INPUT Mark
	    IF Mark > Highest
	      THEN
	         Highest ← Mark
	    ENDIF
	NEXT Counter
	OUTPUT Highest
	``
	
	The pseudocode uses an IF statement to compare the value of each mark with the current value in the variable `Highest`. Whenever the value of the mark is greater than the value in `Highest` the value of mark is assigned to `Highest`. In this way the variable `Highest` maintains a record of the highest mark entered which is output once the loop has completed.
	
	In the above example initialising `Highest` to zero is appropriate as zero is the lowest score that could be achieved. But imagine a situation where the inputs could all be negative numbers. If `Highest` was set to zero in that situation it would remain unaltered as all values input would be less than zero.

(continued)

Task	Description and pseudocode
| Finding maximum | To avoid this situation either initialise `Highest` to a value lower than any value that would be expected `DECLARE Highest : INTEGER ← -10 000`, or alter the code to set the value of `Highest` to the first value input in the series.<br><br>```
DECLARE Highest : INTEGER ← 0
DECLARE Mark : INTEGER ← 0
FOR Counter ← 1 TO 600
    OUTPUT "Enter the mark for the student"
    INPUT Mark
    IF Counter = 1
      THEN
         Highest ← Mark
      ELSE
         IF Mark > Highest
           THEN
              Highest ← Mark
           ENDIF
      ENDIF
NEXT Counter
OUTPUT Highest
``` |
| Finding minimum | The process of finding the lowest (minimum) value of a series of values.<br><br>This is a similar process to finding the maximum value. But it is crucial that the initial value of the variable holding the smallest value is set correctly. It is common to initialise variables with zero, but in most situations, this would be inappropriate for a variable designed to hold the smallest value.<br><br>Consider a system required to output the lowest height of a group of 30 students. The initial value of the variable would need to be larger than the height values input. Selecting an extreme initial value of 3000 centimetres would be appropriate as none of the students will be three metres high.<br><br>```
DECLARE Lowest : REAL ← 3000
DECLARE Height : REAL ← 0
FOR Counter ← 1 TO 30
    OUTPUT "Enter height of student in centimetres"
    INPUT Height
      IF Height < Lowest
        THEN
           Lowest ← Height
        ENDIF
NEXT Counter
OUTPUT Lowest
``` |

(continued)

Task	Description and pseudocode
Finding average	The process of calculating the average value of a series of values.
	To calculate an average value the total value of the series of values is divided by the number of items in the series.
	To find the average of the numbers:
	7, 11, 8, 12, 10
	Find the total of the numbers:
	7 + 11 + 8 + 12 + 10 = 48
	Count how many numbers in the series:
	5 numbers
	Divide the total by the number of values:
	48 / 5 = 9.6
	I am sure you will have spotted that finding an average will involve the use of the code for totalling above.
	Example Task: A system is required that will input the temperature in centigrade for the 31 days in the month of July. The system will output the average temperature for July.
	``
	DECLARE Total: REAL ← 0
	DECLARE Avg : REAL ← 0
	DECLARE Temp : REAL ← 0
	FOR Counter ← 1 TO 31
	OUTPUT "Enter the daily temperature in centigrade"
	INPUT Temp
	//Keeping the total value of all the temperatures input
	Total ← Total + Temp
	NEXT Counter
	//Calculate and output average outside of the loop
	//Divide Total by 31 as 31 temperatures have been input
	Avg ← Total / 31
	OUTPUT Avg
	``

Table 5.4: Common iterative tasks

5.6 Condition-controlled loops

The WHILE...ENDWHILE and the REPEAT...UNTIL loop structures are controlled by a specific condition. Iterations are repeated continuously based on certain criteria. FOR loops are used where the number of iterations is known or can be calculated. WHILE...ENDWHILE and the REPEAT...UNTIL loop structures allow iteration where the number of repetitions is unknown.

Consider the situation where a random number was constantly subtracted from another until the resultant value was less than zero. It would be impossible to determine the number of iterations required to cause the first number to be less than zero. As a result, a FOR loop would not be appropriate; another iterative method would have to be used.

5.7 WHILE loops (pre-condition loops)

In WHILE loops, iterations continue while the loop conditions remain true, regardless of the number of iterations this may generate. The loop will only exit when the loop conditions become false. Two main forms of condition are used.

Condition	Example
The value of user input.	A loop will run while the user inputs a positive number. Inputting a negative number will exit the loop.
Value of a variable used in the code within the loop.	A loop will run while the variable that holds the sum of all the values input is less than 1000. When the value in the variable is greater than 1000 the loop will exit.

Table 5.5: Conditions used in WHILE loops

It is usual for the code within the loop to impact on the values of the condition of the loop in such a way that the criteria will eventually become false and the loop will cease.

Because the conditions are tested at the start (before the loop is run), it is possible that the loop will never run if the conditions are false at that point. It is also possible to accidentally code an **infinite loop** where the conditions remain true for ever.

When writing a **WHILE loop** in Visual Basic, you use the following format:

```
Do While X > 0
    'code to be iterated
    Y = input ("please enter a number")
    X = X - Y
Loop
```

KEY WORDS

infinite loop: a loop that will iterate indefinitely. Usually caused by a logical error in the condition that controls the loop.

WHILE loop: a type of iteration that will repeat a sequence of code while a set of criteria continues to be met. They are also known as pre-condition loops because the condition is checked before the loop executes. If the criteria are not met before the loop starts, the loop will not run and the code within it will not be executed.

Each individual element of the loop performs an important role in achieving the iteration as shown in Table 5.6.

Element	Description
`Do While`	The start of the loop.
`X > 0`	The condition that controls the loop (also called the criteria of the loop). Each time the iteration is run, the condition is evaluated and, if it remains True, the iteration will run. Once the condition is False, execution of the code is directed to the line following `Loop`. In the counter-controlled WHILE loops, it is important that the code that will impact on the criteria of the loop (in this case, X = X − Y) is included within the loop. In this example, the loop will run while X is greater than 0. The code in the loop that impacts the criteria of the loop reduces the value of X by the value of Y (a number input by the user) each time the loop iterates. At some stage, the value of X will reduce below 0 and the loop will exit. If the code that will impact on the criteria of the loop is outside the loop, the condition of the loop will not change and the loop will run forever (an infinite loop).
`Loop`	The end of the current iteration. The program then returns to `Do While` so the condition can be re-evaluated. If the criteria is true, the program will run further iterations. If the criteria is false, the program will break out of the loop and move onto the next section of code.

Table 5.6: Steps in the WHILE loop

DEMO TASK 5.3

A system is required to take a series of positive numbers. The system will keep a running total of the sum of all the numbers and a count of how many numbers have been input. The input of a negative number will indicate the end of the series of numbers. The values of the running total and the number counter will then be output.

Solution

Flowchart 5.3 shows the design of the algorithm. The pseudocode is also shown.

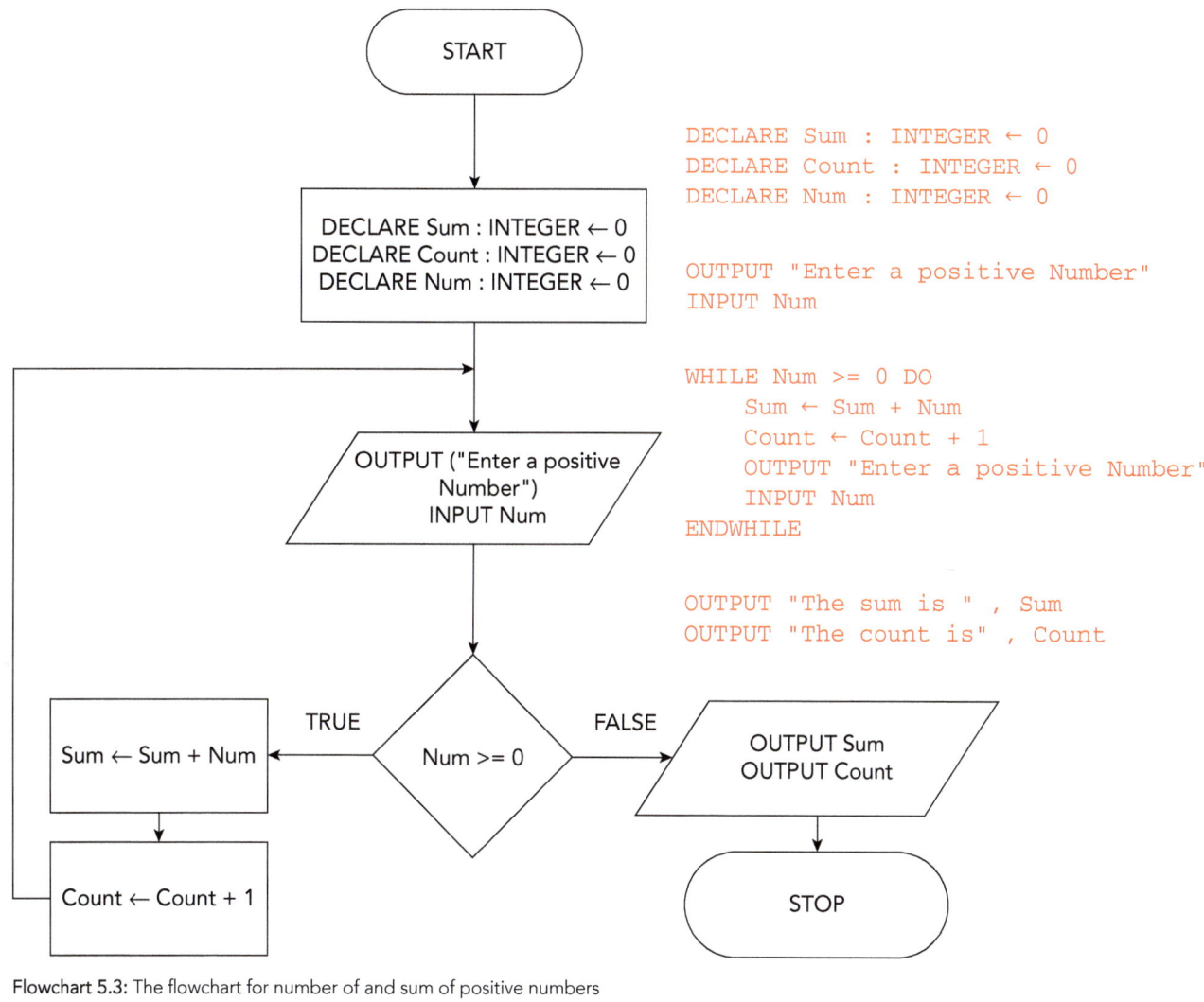

```
DECLARE Sum : INTEGER ← 0
DECLARE Count : INTEGER ← 0
DECLARE Num : INTEGER ← 0

OUTPUT "Enter a positive Number"
INPUT Num

WHILE Num >= 0 DO
    Sum ← Sum + Num
    Count ← Count + 1
    OUTPUT "Enter a positive Number"
    INPUT Num
ENDWHILE

OUTPUT "The sum is " , Sum
OUTPUT "The count is" , Count
```

Flowchart 5.3: The flowchart for number of and sum of positive numbers

> **CONTINUED**
>
> This is the code for a console implementation of this task.
>
> ```vb
>     Module Module1
>         Sub Main()
>             'Declare variables required
>             Dim Sum As Integer = 0
>             Dim Count As Integer = 0
>             Dim Num As Integer = 0
>
>
>             'Get first value for Num - outside of loop
>             'so the WHILE loop has a value to check against criteria
>             Console.WriteLine("Enter a positive number")
>             Num = Console.ReadLine()
>
>
>             'Check Num is positive
>             Do While Num >= 0
>                 'While positive
>                 'Maintain a running total and count of numbers entered
>                 Sum = Sum + Num
>                 Count = Count + 1
>                 'Get subsequent values which are then checked against criteria
>                 Console.WriteLine("Enter a positive number")
>                 Num = Console.ReadLine()
>             Loop
>             'Once loop has exited by input of a negative number
>             'Output the Sum and Counter with appropriate message
>             Console.WriteLine("The sum is " & Sum)
>             Console.WriteLine("The count is " & Count)
>             Console.ReadKey()
>         End Sub
>     End Module
> ```
>
> NOTE: There is a difference between the common pseudocode representation of a WHILE loop and the format of the Visual Basic code used to implement a WHILE loop. While the pseudocode describes the algorithm as `WHILE <condition> DO`, in Visual Basic this is implemented as `Do While <condition>`.

SKILLS FOCUS 5.1

LOOPS

A system is needed to control the weight of cargo crates loaded on to a barge that can only hold 1000 kilograms. As each crate is prepared ready for loading onto the barge, its weight, in kilograms, is entered into a system. The system will keep a running total of the combined weight of the crates being prepared. The system will also check that the combined weight of the crates loaded is not greater than 1000 kilograms. Once this weight is reached, the system will output the number of crates loaded and the combined weight of those crates.

> **CONTINUED**

It is obvious that a **condition-controlled loop** is needed because you are in that section of the book. However, if this task had been given in a test, you would have to identify whether a loop is needed and, if so, what type of loop would be most appropriate.

Step 1 The clues to the need for a loop are contained in the words in the scenario: 'As each crate is prepared' – 'keep a running total' – 'once this weight is reached' – all suggest a repeated process.

Step 2a Consider the type of loop. A FOR loop would not be appropriate as it is not known how many crates would be needed to reach the maximum weight.

A WHILE loop is appropriate as it is important that the weight of each crate is tested *before* being loaded. Checking the combined weight *after* the create has been loaded could mean overloading the barge and require the last crate to be unloaded to bring the weight within range.

Step 2b However, the crate that takes the total weight over 1000 kilograms *will* have been included in the total weight value. This crate will need to be deducted before the total weight on the barge can be output.

Step 3 Consider the criteria needed. In this example, the combined weight must be less than or equal to 1000 kilograms. Look carefully at the wording of the scenario. 'Does not exceed 1000 kilograms' means that 1000 kilograms is acceptable but 1001 kilograms is not. Therefore, the criteria will be <= 1000.

Step 4 Consider the variables needed and their data types. To help identify appropriate variables, consider both the inputs and outputs in the scenario. An example could be `CrateWeight` as an input variable and `CombinedWeight` and `NumberCrates` as output variables.

Step 5 Pseudocode or flowchart the algorithm and check it works as intended.

Reading the task carefully to identify the process required and then selecting appropriate programming constructs to achieve those processes are all important skills for programming. When presented with a task, underline the text to help you identify the required processes.

Tasks that involve iteration are likely to include words such as:

'A number of ..'   'Repeatedly input..'
'All values must be ..' 'Every input will ..'

The next decision you will need to make is the type of loop to use. Consider the features of each type of loop and match those to the task.

Table 5.7 shows the features of different loops.

> **KEY WORD**
>
> **condition-controlled loops:** types of iteration where the repetition of the loop is determined by conditions. The amount of times the loop will be executed is unknown.

CAMBRIDGE IGCSE™ & O LEVEL COMPUTER SCIENCE: PROGRAMMING BOOK

CONTINUED

FOR Loop – Count controlled so loops a known number of times
Indicated by words in the task that quote a number or reference a known value.
'All 20 values will ..'   'There are 60 values ..'   'For each day of the week

WHILE / REPEAT...UNTIL Loop – Condition controlled so that the number of loops is dependent on a condition
Indicated by words in the task that give the criteria.
'Input sequence will be ended with a ...'   'The process will stop when the value is ...'   'When the first match is found ..'

Table 5.7: Features of each type of loop

It is unlikely that tasks will use the words FOR, WHILE or UNTIL, so look for the clues.

Questions

What type of loop might be appropriate for the following?

1. A system used to output the average examination score of a class of 25 students.
2. A system used to record the temperature of a chemical reaction. Temperatures will be recorded every minute. The final record will be when the temperature has reached 20 degrees Celsius.
3. A system used to record the times of 200 marathon runners.
4. A system used in a cinema to record the age range of the people who watch a film. As each person buys a ticket, their age is entered into the system.

PRACTICE TASK 5.3

Consider the system required to load crates onto a barge in Skills Focus 5.1.

- Read through the steps and draw the flowchart or pseudocode solution for your system.
- Test your algorithm works by programming and running the code in Visual Basic.

5.8 WHILE loops with multiple criteria

Often a system needs to be controlled by more than one condition. In the barge-loading task, the barge is likely to have a maximum volume in addition to a maximum weight. Therefore, it might be important to also limit the number of crates as well as the combined weight of the crates.

In the prime number task in Section 5.4, the output was a message indicating if the input number was prime. To achieve this, the FOR loop checked every number between either:

a   2 and the input value minus 1, or
b   2 and half the input value, if you had identified that as a more efficient method.

Consider the situation where the number 120 000 had been input and you were using the second method. Even though the first test of the loop '120 000 Mod 2' had identified that the number couldn't be prime, the FOR loop would still iterate 59 999 times. In this case, 59 998 subsequent iterations are unnecessary.

A WHILE loop can provide a more efficient solution to the prime number task. The WHILE loop would iterate only while no number tested had generated a modulus of zero. The loop can end either at the identification of the first number that generated a modulus of zero or when all relevant integer values have been tested.

To achieve this, the WHILE loop would need two criteria:

- A counter that is used in a similar way to how NEXT is used in the FOR loop. The counter will increment by 1 at each iteration and be used within the code in the same way the loop counter is used in a FOR loop. The value of the counter must be between 2 and half the value of the number being tested.
- A Boolean value that will be initialised as TRUE. The loop will iterate while the value is TRUE. The value will be altered to FALSE when a number tested generates a modulus of zero.

Both criteria must be true for the loop to continue. Flowchart 5.4 shows the flowchart for the design of the algorithm. The pseudocode is also shown.

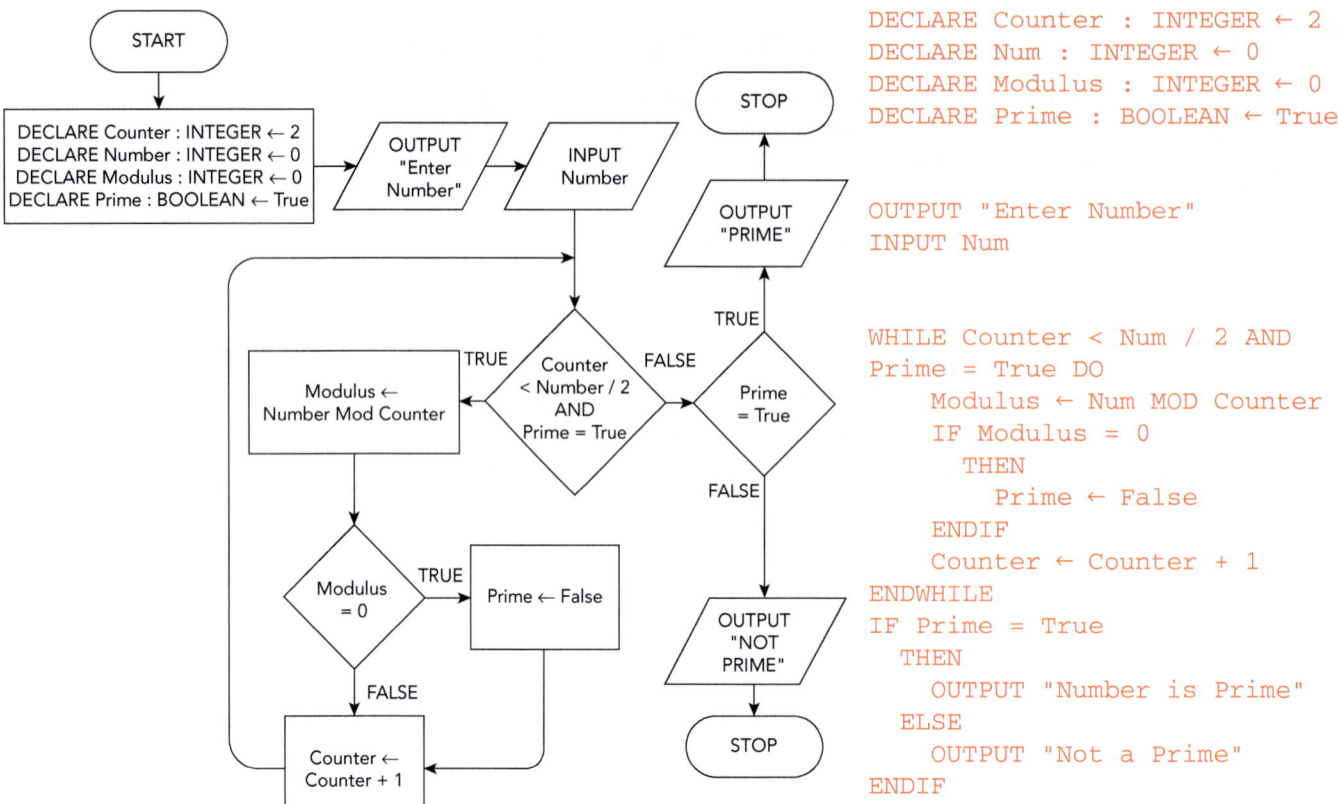

Flowchart 5.4: Flowchart for WHILE approach

```
DECLARE Counter : INTEGER ← 2
DECLARE Num : INTEGER ← 0
DECLARE Modulus : INTEGER ← 0
DECLARE Prime : BOOLEAN ← True

OUTPUT "Enter Number"
INPUT Num

WHILE Counter < Num / 2 AND
Prime = True DO
    Modulus ← Num MOD Counter
    IF Modulus = 0
      THEN
        Prime ← False
    ENDIF
    Counter ← Counter + 1
ENDWHILE
IF Prime = True
  THEN
    OUTPUT "Number is Prime"
  ELSE
    OUTPUT "Not a Prime"
ENDIF
```

The following is the code for a console implementation:

```vbnet
Module Module1
    Sub Main()
        Dim Num As Integer = 0
        Dim Modulus As Integer = 0
        Dim Prime As Boolean = True
        'Obtain and store input from user
        Console.WriteLine("Insert Number to Test")
        Num = Console.ReadLine()
        'Declare counter to use with the While loop
        'Initialise to 2 to avoid using 1 as divisor
        Dim Counter As Integer = 2
        'Start the While loop with two conditions
        'AND connecting operator means BOTH conditions must be
        True to iterate
        Do While Counter < Num / 2 And Prime = True
            Modulus = Num Mod Counter
            'If statement changes Prime to False if modulus = 0
            'This will cause the loop conditions to be

            'False at next iteration and stop loop
            If Modulus = 0 Then
                Prime = False
```

```vb
            End If
            'Counter incremented to loop through all
            'integer values between 2 num / 2
            Counter = Counter + 1
        Loop
        'Once loop complete
        'If statement outputs appropriate message based on
            'state of Prime Boolean
        If Prime = True Then
            Console.WriteLine("PRIME number")
        Else
            Console.WriteLine("NOT a PRIME number")
        End If
        Console.ReadKey()
    End Sub
End Module
```

> ### PRACTICE TASK 5.4
>
> A system is required that will calculate the average value from a series of positive integer values input by the user with the following criteria:
>
> - The system will accept a maximum of 20 values.
>
> - The series of input values can be terminated before all 20 values have been entered by entering the value −1.
>
> The system will calculate and output the average of values input.
>
> Design the pseudocode and create the system. Test your system to ensure it works as expected.

> ### CHALLENGE TASK 5.2
>
> Design and create a system that will take two positive integer values as input. The system will output all the common factors of the numbers input. A factor is a number that will divide into both numbers without leaving a remainder.

> ### TIP
>
> Loop to find if a value is a factor of the first number. Then check if the value is also a factor of the second number. Only if the value divides into both numbers without a remainder will it be a common factor.

5.9 REPEAT...UNTIL loops (post-condition loops)

A **REPEAT...UNTIL loop** is very similar to a WHILE loop as iteration will continue based on the loop conditions. Like the WHILE loops, it is also able to work in situations where the number of iterations is unknown.

Unlike in a WHILE loop, the test is completed at the end of the iteration so the iterated code will always run at least once. REPEAT...UNTIL loops are also known as post-condition loops because the condition is checked at the end of the loop.

When writing a REPEAT...UNTIL loop in Visual Basic, you need to use the following format.

```
Do
    'Code for iteration
    Y = Input ("please enter a number")
    X = X - Y


Loop Until X <= 0
```

> **KEY WORD**
>
> **REPEAT...UNTIL loop:** a type of iteration that will repeat a sequence of code until a certain condition is met. The code within the loop will always be executed at least once. They are also known as post-condition loops.

The individual elements of the code perform an important role in the iteration, as shown in Table 5.8:

Element	Description
`Do`	The start of the loop. At every iteration, the execution of the program will be passed to the `Do` command. Because the loop starts before any conditions are checked, the iteration will always run at least once.
`Loop Until`	The end of the loop.
`X <= 0`	The condition for the loop. Each time the iteration is run, the condition is evaluated. If it remains False, the execution is directed to `Do` and the iteration will run again. Once the condition is True, the execution of the code is directed to the line following `Loop Until`. It is possible to use the logical operators (AND, OR and NOT) if you want to use multiple conditions. Counter-based conditions require the counter to incremented by the code contained within the loop.

Table 5.8: Elements of the REPEAT...UNTIL loop

To compare the way in which WHILE and REPEAT...UNTIL loops, operate consider the code in Table 5.9:

WHILE	REPEAT...UNTIL
| ```
Do While X > 0
    Y = input ("enter a number")
    X = X - Y
Loop
Output X
``` | ```
Do
    Y = Input ("enter a number")
    X = X - Y
Loop Until X <= 0
Output X
``` |

Table 5.9: Comparing the WHILE and REPEAT...UNTIL loops

Table 5.10 shows the output from both loops when the numbers entered by the user are in the order 10, 20, 30 and 20:

Starting value of X	Output from WHILE loop	Output from REPEAT...UNTIL loop
0	Output 0 The check is at the start of the loop. X is not greater than zero, so the criteria is false. The loop will not run and X remains at zero.	−10 The code in the loop runs once: X = 0 − 10 so X becomes −10 The criteria is then checked and, as X is less than zero, the criteria is true and the loop will exit.
80	Output 0 The loop would accept the values 10, 20, 30 and 20 as inputs. The final input would reduce the value of X to zero. The criteria when checked would be false and the loop would exit.	Output 0 The loop would accept the values 10, 20, 30 and 20 as inputs. The final input would reduce the value of X to zero. The criteria when checked would be true and the loop would exit.

Table 5.10: Output of loops

- A REPEAT...UNTIL loop always runs once, irrespective of the criteria check.
- The criteria in the two loops operate very differently. As a result, selecting the criteria values is crucial.

DEMO TASK 5.4

The algorithm to calculate if a number is prime in Section 5.4 can also be coded using a REPEAT...UNTIL loop.

Solution

Flowchart 5.5 shows the REPEAT...UNTIL loop. The pseaudocode is also shown. If you compare these with the WHILE approach in Flowchart 5.4, you will be able to identify the differences in approach.

The decision criteria are checked at different stages during the process: the WHILE at the start of the loop and the REPEAT...UNTIL at the end. The WHILE loop runs if the criteria is TRUE and the REPEAT...UNTIL loops if the criteria is FALSE.

CONTINUED

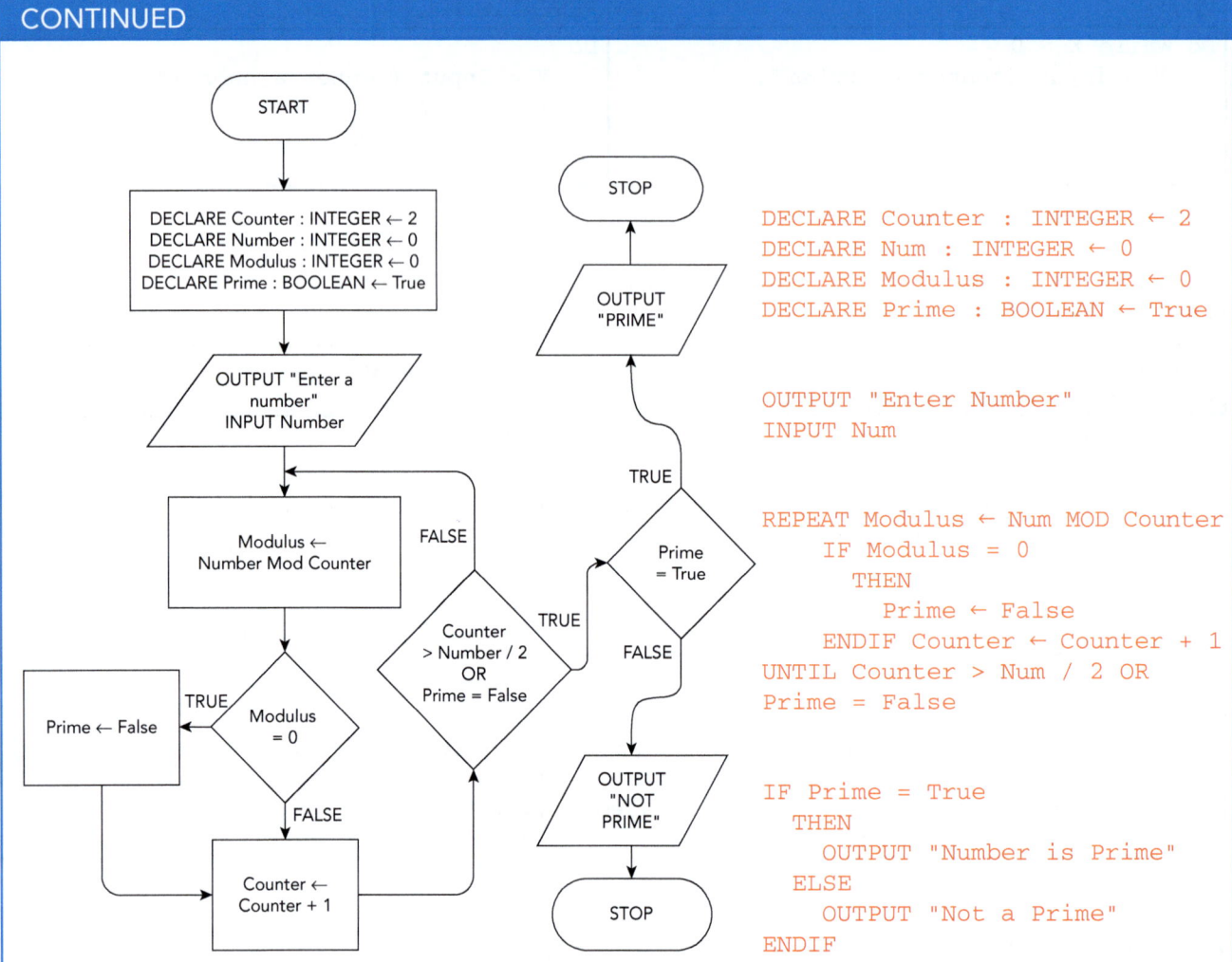

Flowchart 5.5: Flowchart from prime number testing using REPEAT...UNTIL

The following is the code for a console application implementation:

```
Module Module1
    Sub Main()
        Dim Prime As Boolean = True
        Dim Num As Integer = 0


        Console.WriteLine("Input NUMBER to Test")
        Num = Console.ReadLine()


        'Declare counter to use with the UNTIL loop
        'Initialise to 2 to avoid using 1 as divisor
        Dim Counter As Integer = 2
```

> **CONTINUED**
>
> ```vb
>             'Start the Until loop with two conditions that are checked at end
>         Do
>             'IF statement changes Prime to False if modulus = 0
>             'This will cause the loop conditions to be False at next check and
>             stop loop
>             If Num Mod Counter = 0 Then
>                 Prime = False
>             End If
>             'Counter incremented to loop through all integer values between 2
>             and num - 1
>             Counter = Counter + 1
>             'Loop Until indicates end of current iteration; conditions are
>             checked
>             'The OR connecting operator means EITHER condition could be True
>             to end
>             'Both conditions must be False for execution to be passed to the DO
>         Loop Until Counter = Num - 1 Or Prime = False
>
>         If Prime = True Then
>             Console.WriteLine("PRIME number")
>         Else
>             Console.WriteLine("NOT a PRIME number")
>         End If
>         Console.ReadKey()
>     End Sub
> End Module
> ```

5.10 WHILE and REPEAT...UNTIL loops based on user input

Both WHILE and REPEAT...UNTIL loops are capable of working in a program or section of code where a user will input a sequence of data items and the end of the loop is determined by either the inputting of a specific item or the value of the inputs reaching a certain value. An example of this approach has already been given in this chapter in the 'barge loading' task in Section 5.6 where the loop finished when the weight exceeded 1000 kilograms.

When using iteration based on user input, it is crucial that the input is included within the loop. A common error is to include a single input outside the loop.

```
OUTPUT "Enter a number"
INPUT Num
WHILE Num > 5 DO
    OUTPUT Num
ENDWHILE
```

Consider the situation if the user input the number 10. The loop will continually check against the value of 10 and run an infinite number of times as Num will always be greater than 5. This is known as an infinite loop.

> ### DEMO TASK 5.5
>
> A system is required that will allow a user to input a series of positive numbers and indicate the end of the sequence by inputting a value of −1. The system will output the sum of the positive numbers input.
>
> Solution
>
> Flowchart 5.6 shows a WHILE and REPEAT...UNTIL loop.
>
>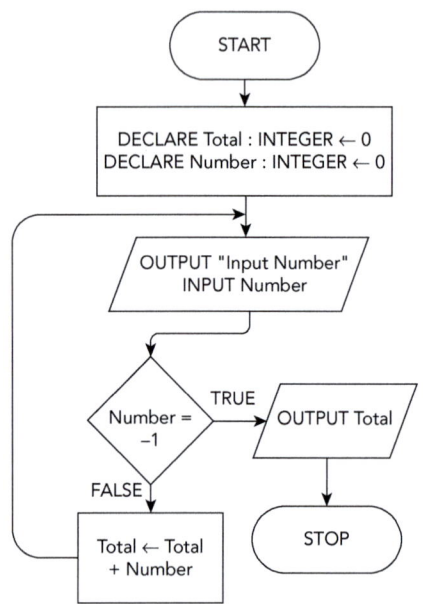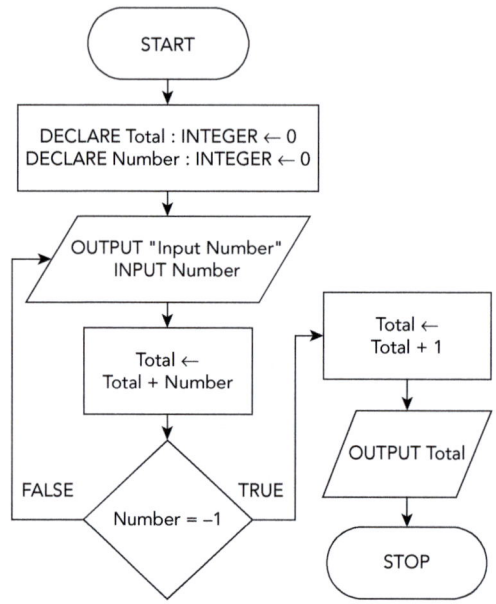
>
> Flowchart 5.6: Flowchart for an input loop using a WHILE and REPEAT...UNTIL loop
>
> Pseudocode for WHILE loop
>
> ```
> DECLARE Total : INTEGER ← 0
> DECLARE Number : INTEGER ← 0
>
> INPUT Number
>
> WHILE Number is not -1 DO
>     Total ← Total + 1
>     INPUT Number
> ENDWHILE
>
> OUTPUT Total
> ```
>
> Pseudocode for REPEAT...UNTIL loop
>
> ```
> DECLARE Total : INTEGER ← 0
> DECLARE Number : INTEGER ← 0
>
> REPEAT
>     INPUT Number
>     Total ← Total + Number
> UNTIL Number = -1
>
> Total ← Total + 1
>
> OUTPUT Total
> ```

5 Iteration

> **CONTINUED**
>
> The different approaches can be seen in the structure of the flowchart and the pseudocode:
>
> - The WHILE loop requires the first number to be input outside the loop to provide a value to check.
> - The REPEAT...UNTIL loop will have included the input of −1 in `Total` as the conditions are not checked until after the processing in the loop has been completed. As a result, the total has to be recalculated after the loop.
>
> Code snippet 5.1 shows the code for a console application using a WHILE loop and a REPEAT...UNTIL loop.
>
> **WHILE code**
>
> ```vb
> Module Module1
>     'Declare and initialise variables
>     Dim Number As Integer = 0
>     Dim Total As Integer = 0
>
>     Sub Main()
>         'Output to screen input instructions
>         Console.WriteLine("Please input a Number")
>         Console.WriteLine("Input -1 to End Input Sequence")
>
>         'Obtain first input value
>         Number = Console.ReadLine()
>
>         'Start loop and check if input is -1
>         Do While Number <> -1
>
>             'Increment Total to maintain a sum of numbers input
>             Total = Total + Number
>
>             'Output instructions and obtain next input value
>             Console.WriteLine("Please input a Number")
>             Console.WriteLine("Input -1 to End Input Sequence")
>             Number = Console.ReadLine()
>             'Indicate end of loop and redirects execution to
>             'start of the WHILE loop
>         Loop
>
>         'Output the value within the variable Total
>         Console.WriteLine(Total)
>         'Used to hold Total in display until a key pressed
>         Console.ReadKey()
>
>     End Sub
>
> End Module
> ```

CONTINUED

REPEAT...UNTIL code

```vb
Module Module1
    'Declare and initialise variables
    Dim Number As Integer = 0
    Dim Total As Integer = 0


    Sub Main()


        Do
            'Output instructions and obtain input value
            Console.WriteLine("Please input a Number")
            Console.WriteLine("Input -1 to End Input Sequence")
            Number = Console.ReadLine()


            'Increment Total to maintain sum of numbers
            Total = Total + Number


            'Indicate end of loop and redirects execution
            'to start of the UNTIL loop
        Loop Until Number = -1


        'Increase Total by 1 as the -1 input was included
        'in Total because the condition is checked
        'after the processing
        Total = Total + 1


        'Output the value within the variable Total
        Console.WriteLine(Total)
        'Used to hold Total in display until a key pressed
        Console.ReadKey()


    End Sub


End Module
```

Code snippet 5.1: Code for console application using WHILE and REPEAT...UNTIL

5.11 Choosing to use WHILE or UNTIL

As can be seen from Demo Task 5.5, both loops can perform the same task. So what might influence your choice of loop for any given task?

To make this decision, we need to consider the key difference between the two loops. These are shown in Table 5.11.

Loop	Difference
WHILE	Will check criteria at the start of the loop. As a result, the loop may not run.
REPEAT...UNTIL	Will check criteria at end of loop. As a result, the loop will always run at least once.

Table 5.11: Differences between WHILE and REPEAT...UNTIL loops

WHILE loops are therefore most appropriate in tasks where an input needs to be tested to decide if the loop should run. Tasks where a series of inputs are totalled or compared with a particular input value are suited to the use of a WHILE loop. Because the input is tested at the start of loop, it avoids the end value being included in the totalling or comparison.

REPEAT...UNTIL loops are most appropriate in tasks where the process within the loop must run at least once. Or where the criteria to be checked are generated within the loop.

> **PRACTICE TASK 5.5**
>
> Consider which approach might be most appropriate in these scenarios:
>
> 1. A program used to total a series of inputs. The series of inputs will end when a negative number is input.
> 2. A program used to stop a series of numerical inputs as soon as the total of those inputs is above 1000.
> 3. A program that allows user access to a system if the username and password input match the system records. The system will reject incorrect inputs and prompt for re-entry.

5.12 Nested iteration

In Chapter 4 we looked at nested selection, where one IF statement is contained (nested) entirely with another. This process allowed for multiple decisions to be made, the first decision leading to subsequent decisions.

A similar process can be followed using iteration. Consider the situation where a user is required to input two different sets of five numbers, each set will be totalled, and the two totals will be output. This could be achieved by running two individual FOR loops as shown in the following pseudocode. Each loop accepts and totals five input values before outputting the total of those five values.

```
DECLARE Number : REAL ← 0
DECLARE Total : REAL ← 0
FOR Counter1 ← 1 TO 5
    OUTPUT "Input a number"
    INPUT Number
    Total ← Total + Number
NEXT Counter1
OUTPUT Total
TOTAL ← 0
FOR Counter2 ← 1 TO 5
    OUTPUT "Input a number"
    INPUT Number
    Total ← Total + Number
NEXT Counter2
OUTPUT Total
OUTPUT "All complete"
```

An alternative approach could make use of nested iteration. One FOR loop (highlighted in yellow) used to take and total five inputs before outputting the total. This loop would be contained (nested) inside another loop that would cause the totalling loop to run twice.

```
1. DECLARE Number : REAL ← 0
2. FOR Counter1 ← 1 TO 2
3.     DECLARE Total : REAL ← 0
4.     FOR Counter2 ← 1 TO 5
5.         OUTPUT "Input a number"
6.         INPUT Number
7.         Total ← Total + Number
8.     NEXT Counter2
9.     OUTPUT Total
10. NEXT Counter1
11. OUTPUT "All complete"
```

Table 5.12 shows the execution of this code.

Step	Line number(s)	First loop	Output
1	2, 3	The first loop iterates for the first time. Value of `Counter1` is 1. `Total` is declared and initialised as 0.	
2	4, 5, 6, 7, 8 Will be iterated five times.	The second loop will run five times. At each iteration, the user will be prompted to input a value which is then totalled in the variable `Total`. The value of `Counter2` will be incremented at each iteration until it exceeds 5, at which time the loop will exit. Five values will have been input and totalled.	

(continued)

Step	Line number(s)	First loop	Output
3	9	The value of `Total` is output.	The total of the first five values input.
4	10	The NEXT instruction of the first loop is reached so `Counter1` is incremented to become 2.	
5	2, 3	The first loop Iterates the first time. Value of `Counter1` is 2. `Total` is declared and initialised as 0, deleting all the earlier values.	
6	4, 5, 6, 7, 8 Will be iterated five times.	The second loop will be run again. `Counter2` will start from a value of 1 and the loop will therefore run another five times. At each iteration, the user will be prompted to input a value which is then totalled in the variable `Total`. The value of `Counter2` will be incremented at each iteration until it exceeds 5, at which time the loop will exit. Another five values will have been input and totalled.	
7	9	The value of `Total` is output.	The total of the second five values input.
8	10	The NEXT instruction of the first loop is reached so `Counter1` is incremented to become 3.	
9	2	As the value of `Counter1` is beyond the condition limit the loop will end.	
10	11	The first loop will exit and line 11 is executed.	"All complete"

Table 5.12: Execution of nested iteration code

In this example the number of lines of code in the nested iteration example is similar to the simple individual loop example. Consider the situation if the input of sets of five numbers had to be repeated 100 times.

Uses of nested iteration

Nested iteration can be effective where a repetitive process needs to be repeated more than once. It is particularly effective with grid like data formats.

For example, a monitor consists of thousands of pixels arranged in rows. Each row will contain many pixels. The code to turn the entire screen a single colour could be written using nested iteration.

```
FOR each row on the monitor
    FOR each pixel in the row
        Change pixel to required colour
    NEXT Pixel
NEXT Row
```

Later in the book (Chapter 10) you will learn about arrays. Arrays are data structures that can hold multiple data items under a single identifier (name). You will discover how nested iteration can be used to sort and search these types of data structure.

SUMMARY

Iteration provides methods that programmers can use to loop through sequences of code multiple times. There are three basic forms of iteration: FOR loops, WHILE loops and REPEAT...UNTIL loops.
Loops in flowcharts are shown through the use of a decision element (diamond shape) with a flow line looping back to an earlier element of the diagram. The decision element contains the criteria on which the iteration is based.
A FOR...TO...NEXT loop is used where the number of iterations is known at the outset.
A FOR loop contains a loop counter to record the number of iterations. They are known as count-controlled loops. This value is checked each time the loop executes. This value can be used within the code in the FOR loop.
Condition-controlled loop structures, such as WHILE...DO...ENDWHILE or REPEAT...UNTIL, are used where the number of iterations is unknowns.
WHILE...DO...ENDWHILE structures check the loop conditions at the outset of the loop. They are known as pre-conditiion loops. If the conditions are False, the loop will never run.
REPEAT...UNTIL structures check the loop conditions at the end of the first iteration. They are known as post-condition loops. The loop will always run at least once.
Nested iteration is when one loop is contained entirely within another loop. It is used to run an iterative process more than once.

END-OF-CHAPTER TASKS

For each task, you should draw a flowchart and create a pseudocode algorithm before programming and running the code in Visual Basic.

1 Design and create a system that will take a positive integer value as input. The system will output all the factors (a number that will divide into the number without a remainder) of that number.

2 Design and create a system that will take a series of positive numerical values greater than zero. The user will be repeatedly prompted to input numerical values until they enter the value zero to end the input stream. The system will then output:
   a The mean average of all the values input.
   b An adjusted average. The system will reject the largest value input and then calculate the mean average of the remaining values.

   For example, if the input values are 20.5, 10, 35, 14.5, and 20. The value 35 is rejected by the system and the average calculated as (20.5 + 10 + 14.5 + 20) / 4 = 16.25.

CONTINUED

3 a Design and create a system that completes a verification check on a password. The system prompts the user to input their password twice. The system will then check if both the inputs are identical. If they are identical, the system will output the message 'Accepted' and the system will exit. If the password inputs are not identical, the user will be prompted to input the password again. This process will continue if the passwords are not identical when re-entered.

b Extend the password verification system to only allow a user to have three attempts at entering their password. At the third failed attempt, the system will output the message 'Account Locked' and the system will exit.

4 a A sugar producer requires a system to check that bags of sugar meet minimum weight requirements. The bags are produced in batches of 120 and 10% of each batch is checked. If the bags in the 10% sample are all within tolerance, the batch is passed. If any of the bags in the 10% sample fail the check, the batch is rejected and the sugar is repacked.

Each bag of sugar should weight 2 kilograms. To pass the check, the bags must weigh at least 2 kilograms and cannot be greater than 10% overweight.

  i Using pseudocode or a flowchart, design and create a system that will input the weight of each of the bags in the 10% sample. The system will check the bag in the sample and output a message to indicate if the batch is accepted or rejected.

  ii Create and test your system using a console application.

> **TIP**
>
> You will need to consider the inputs and outputs involved, the variables required and the most appropriate type of loop.

b The sugar producer also produces bags of sugar weighing 0.5 kilograms and 5 kilograms. The tolerance for these bags are shown in the table:

Weight of Bag	Minimum Weight	Maximum Overweight Limit
0.5 kilogram	0.5 kilograms	20%
5 kilograms	5 kilograms	5%

Extend your system so that the type of bags being checked can be input at the start of the process. The system will check the sample using the appropriate tolerance for the type of bag being checked.

Chapter 6
Designing algorithms

IN THIS CHAPTER YOU WILL:

- learn that systems are made up of subsystems, which in turn may be made up of further subsystems
- understand how to apply top-down design and structure diagrams to simplify a complex system
- combine sequence, selection and iteration to design complex systems
- understand how to produce effective and efficient solutions to complex tasks.

Introduction

If you have a problem, you can often find the solution through logical thinking, identifying the individual elements of the problem and organising a structured plan to complete them. By overcoming the smaller issues, you can overcome the whole problem.

When designing algorithms, a similar process should take place. The process requires the ability to analyse a scenario-based task, identify the individual elements of the task and use programming concepts to create an appropriate algorithm. This is known as computational thinking. Often more than one approach to a given scenario will produce a working solution, which makes this process exciting. Identifying and designing an efficient solution to a problem is at the heart of computational thinking and good programming.

6.1 Top-down design

Top-down design breaks a large problem into simple steps or tasks. Each of these tasks may be split into a number of smaller subtasks. The process is complete once the problem has been broken down sufficiently to allow it to be understood and programmed. This process is also known as 'step-wise refinement'. Breaking up complex problems into smaller, more manageable parts is an important **computational thinking** skill called **decomposition**.

The main advantage of designing solutions in this way is that the final process will be well structured and easier to understand. It can increase the speed of development as different subtasks can be given to individual members of a programming team. This design approach also helps when debugging or modifying programs. Changes can be made to individual subtasks without necessarily having to change the overall program.

This approach is effective in solving large, real-world problems as well as the type of scenario-based questions.

6.2 Structure diagrams

One tool available to programmers when trying to design a solution to a more complex system is the **structure diagram**. When using top-down design principles, the aim is to decompose the system into smaller and smaller problems until no further decomposition can take place. At this point, the programmer can develop algorithms for each sub-problem in the usual way using flowcharts or pseudocode. Structure diagrams make use of diagrams to represent the programming constructs of sequence, selection and iteration. These diagrams help design and refine a task into a series of subtasks. The diagrams use the symbols shown in Table 6.1.

> **KEY WORDS**
>
> **top-down design:** a way of designing a computer program by breaking down the problem into smaller problems (subsystems) until it is sufficiently defined to allow it to be understood and programmed.
>
> **computational thinking:** the thought processes involved in expressing solutions as computational steps or algorithms that can be carried out by a computer. These include abstraction, problem analysis, step-wise refinement, decomposition and algorithm design.
>
> **decomposition:** a computational thinking skill that involves thinking about large tasks and breaking them down into smaller tasks.
>
> **structure diagrams:** a method of expressing a system as a series of subsystems using a diagram.

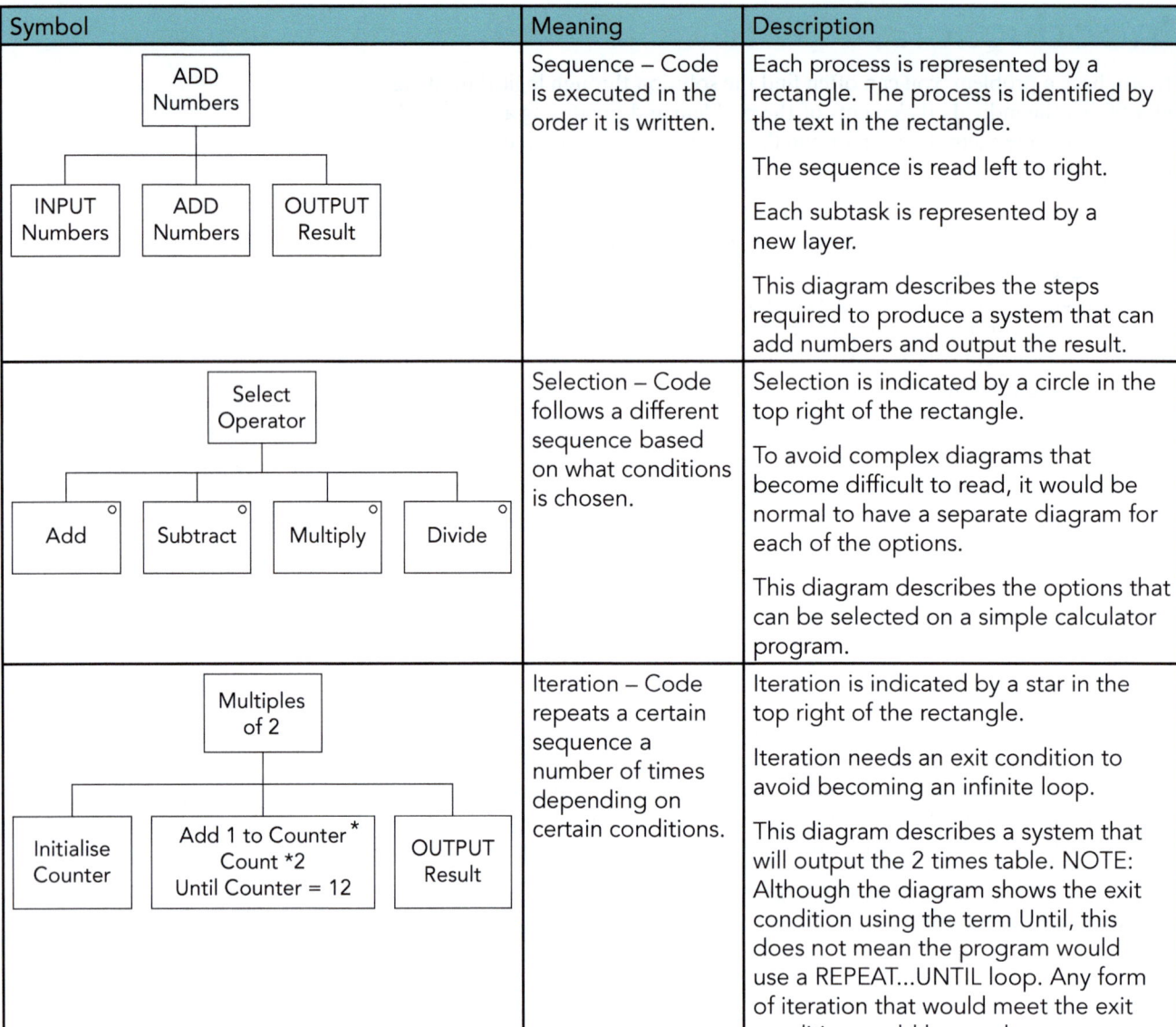

Table 6.1: Sequence, selection and iteration symbols in a structure diagram

> ### DEMO TASK 6.1
>
> *A school requires a system that will check if students are present in lessons and report to parents any absence. The teacher will record if students are in the lesson and then transfer the record to the school administration team. They will contact parents if students are absent. The process of reporting to parents can be done either by telephone or email.*
>
> ### Solution
>
> One possible strategy is to consider the situation in terms of the steps of a computer model:
>
> $$\text{Input} \rightarrow \text{Storage} \rightarrow \text{Process} \rightarrow \text{Output}$$

CONTINUED

We could start the design by considering the main elements of the required system:

- Input: The main input will be the record of the students' presence.
- Storage: To store this, some form of list could be used.
- Process: Recording the presence of every student who attends the lesson.
- Output: Admin will then telephone or email parents.

Figure 6.1 shows the initial structure diagram for this system:

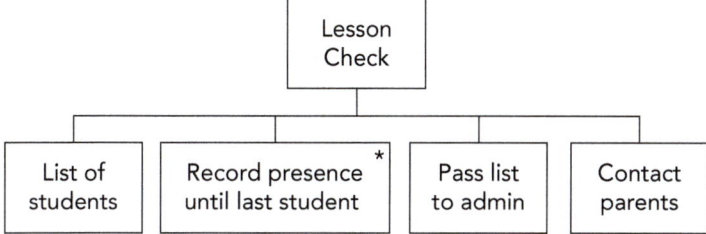

Figure 6.1: Initial structure diagram

The next step is to consider if any of the tasks could be broken down into subtasks. Creating the 'List of Students' is a high-level task that could be broken down into subtasks. Contacting parents can be done by telephone or email and will therefore also need subtasks. Figure 6.2 shows the amended structure diagram with these subtasks.

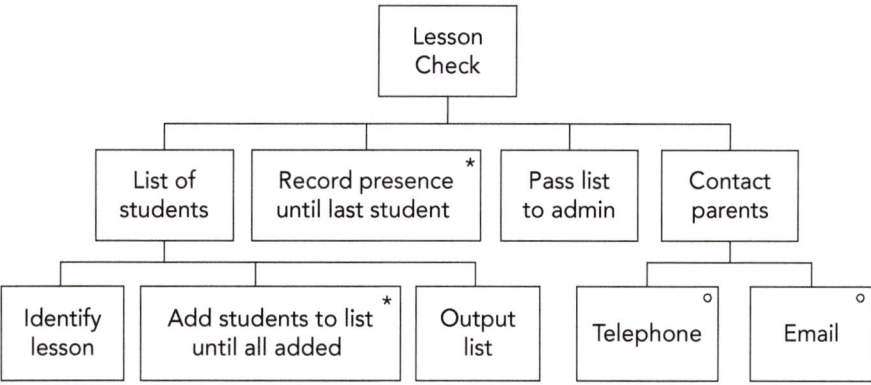

Figure 6.2: Structure diagram showing subtasks

The diagram may not yet be complete. The process 'Identify Lesson' could be broken into further subtasks. What inputs would be needed? Where is the data about which students attend the lesson stored? How will the system access those records? This and other tasks would require consideration if this was a real-life scenario. This is continued until all subtasks have been identified.

> **CAMBRIDGE IGCSE™ & O LEVEL COMPUTER SCIENCE: PROGRAMMING BOOK**

PRACTICE TASK 6.1

Weather app

A weather app works by using location data entered by the user, either a new town or a chosen town from a previously saved location. The weather app will then output the day's forecast either as a visual map or a table of temperatures, wind speeds and weather icons. A weather app is an example of a computer system that is made up of subsystems. The structure diagram in Figure 6.3 shows some of its subsystems. Complete the diagram by filling in the empty boxes:

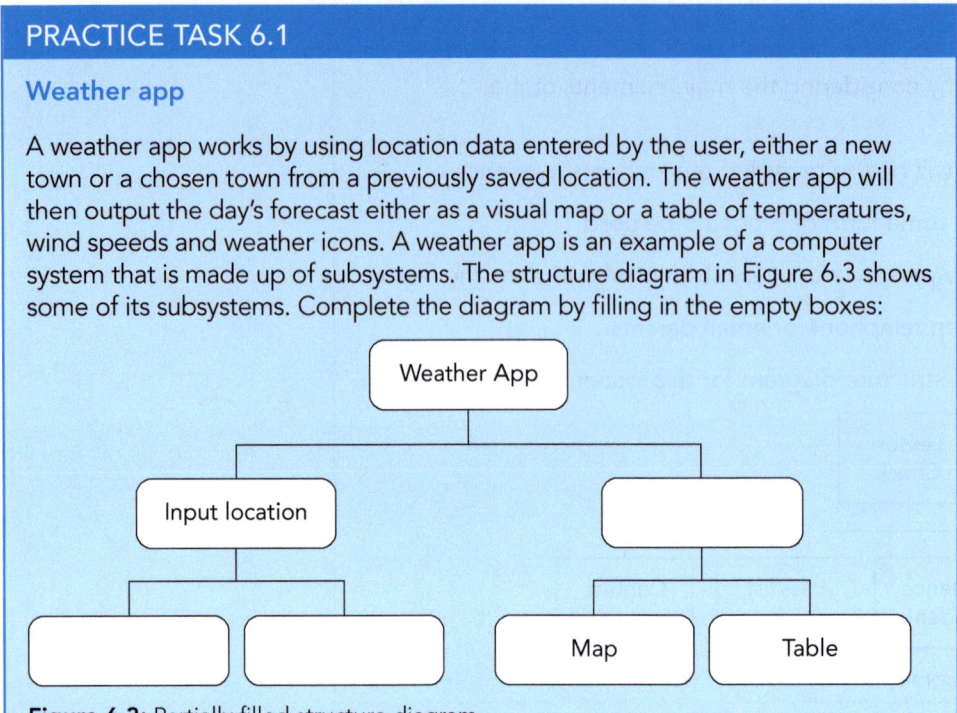

Figure 6.3: Partially filled structure diagram

6.3 Design steps

As suggested in Section 6.2, splitting the overall task into subtasks can be done following the Input → Storage → Process → Output computer model.

When designing an algorithm for a subtask, it is recommended that you make a slight adaption to the process:

1. Identify the inputs and outputs that are involved in the scenario. At this stage, it is worth identifying any global variables that will be required.
2. For each input, identify if the task requires this input to be repeated. This will mean some form of iteration is needed. Identify the most appropriate loop to use.
3. For each output, identify the required calculation or recording process required to produce the output value.
    a. Does the process involve any decision making? This could mean use of an IF statement.
    b. Does the process involve repeated calculation? This could mean some form of iteration.
4. Consider the sequence in which the various processes need to be completed:
    - Check that inputs or processes that need to be iterated are within the loop.
    - Check that single inputs and outputs are outside the loop.
    - Check that all iterations repeat and exit as expected.
    - Check you have defined and initialised the variables or constants that are to be used.

> **TIP**
>
> When programming, your IDE will alert you to incorrect statements and missing variables. When designing in pseudocode, this support is not available. Always check that you have initialised all global variables correctly and that program statements are complete.

Considering the inputs and outputs at the start will help you to consider the aim of the system. If you don't consider the required outputs early in the design stage, how can you define the process required to produce the outputs?

Although you consider variables at the outset, you can finalise your variables as a last step to making sure nothing has been missed.

> ### DEMO TASK 6.2
>
> *A user will input 100 positive numbers into a system. The system will output the highest number and the total sum of the numbers input. Design an appropriate algorithm.*
>
> Solution
>
> Figure 6.4 outlines the computational thinking process for designing the algorithm using the steps described in Section 6.3.
>
>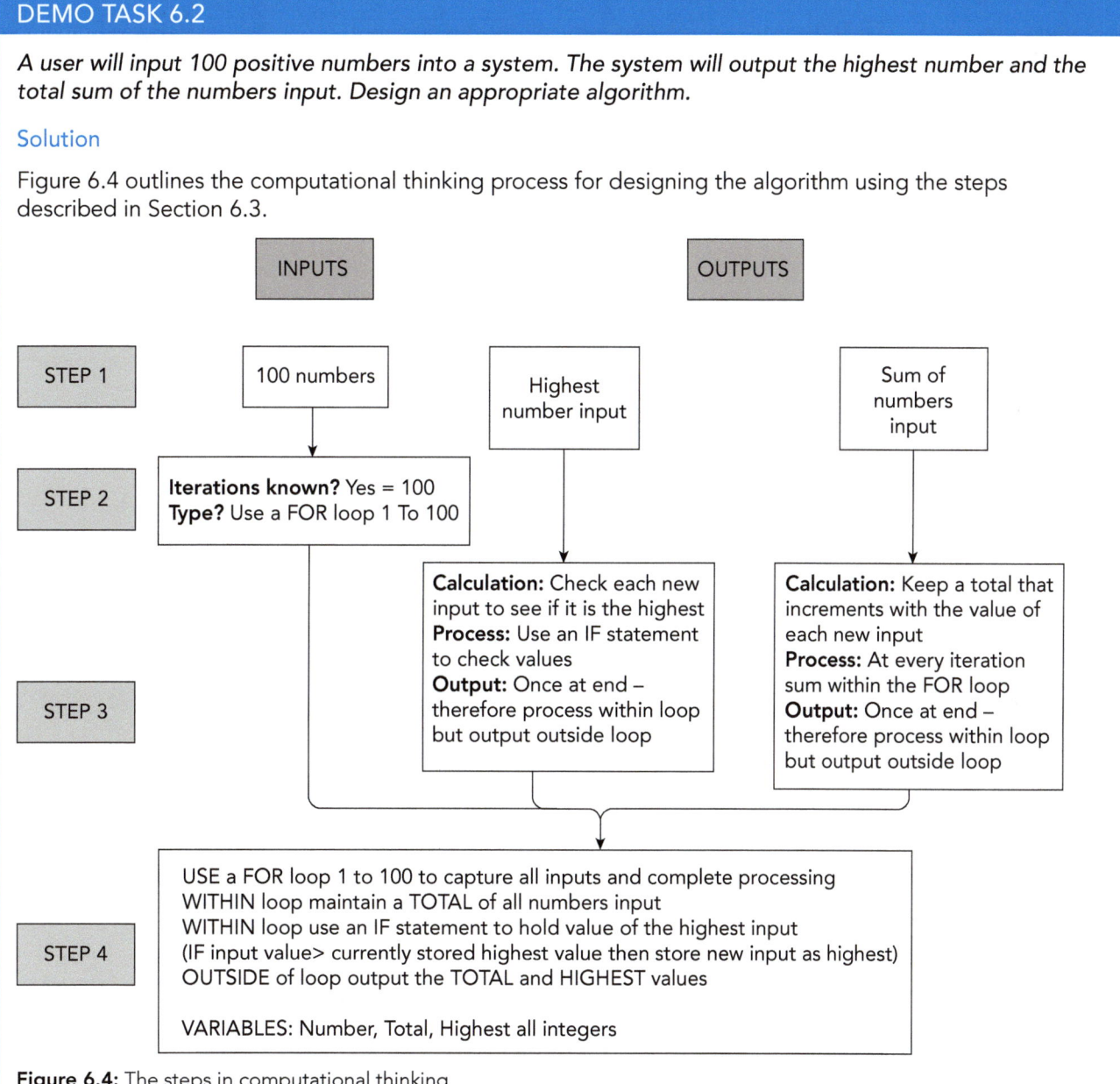
>
> Figure 6.4: The steps in computational thinking

> # CONTINUED

The structure diagram in Figure 6.5 shows the design of the process.

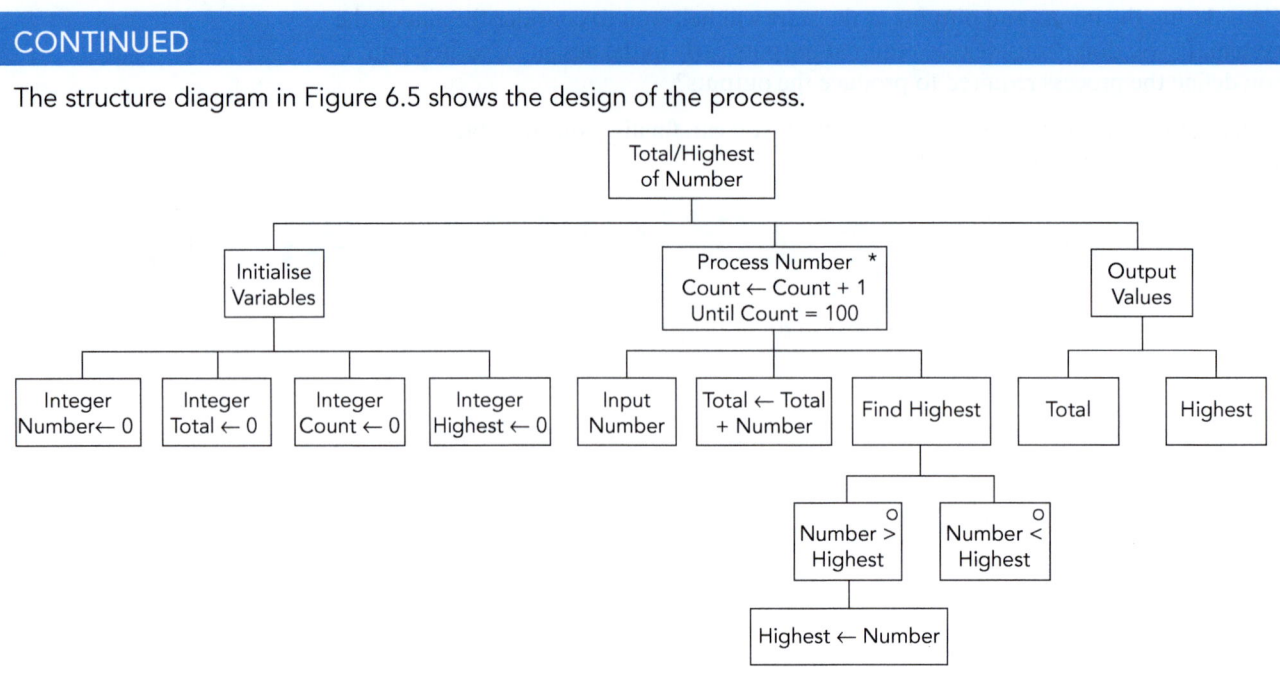

Figure 6.5: Structure diagram showing the design of the process

From this, we can produce the flowchart, shown in Flowchart 6.1, and the pseudocode.

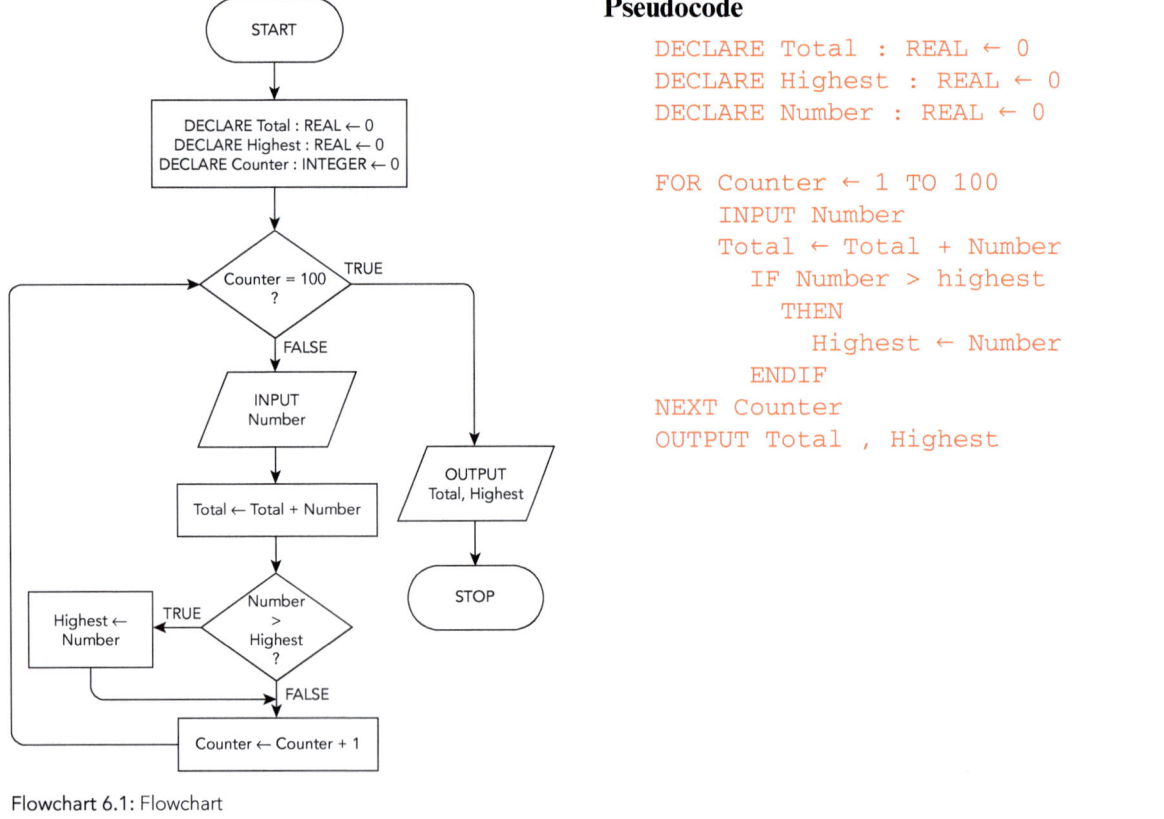

Pseudocode

```
DECLARE Total : REAL ← 0
DECLARE Highest : REAL ← 0
DECLARE Number : REAL ← 0

FOR Counter ← 1 TO 100
    INPUT Number
    Total ← Total + Number
      IF Number > highest
        THEN
            Highest ← Number
        ENDIF
NEXT Counter
OUTPUT Total , Highest
```

Flowchart 6.1: Flowchart

6 Designing algorithms

PRACTICE TASK 6.2

Discussion question: The solution shown in Demo Task 6.2 is not the only acceptable solution. Identify two other ways this algorithm could have been written.

DEMO TASK 6.3

A user is required to input the population of a number of local cities. They will indicate the end of the input sequence by inputting a negative value. The system will output the average population of the cities input.

Solution

Figure 6.6 outlines the computational thinking process for designing the algorithm.

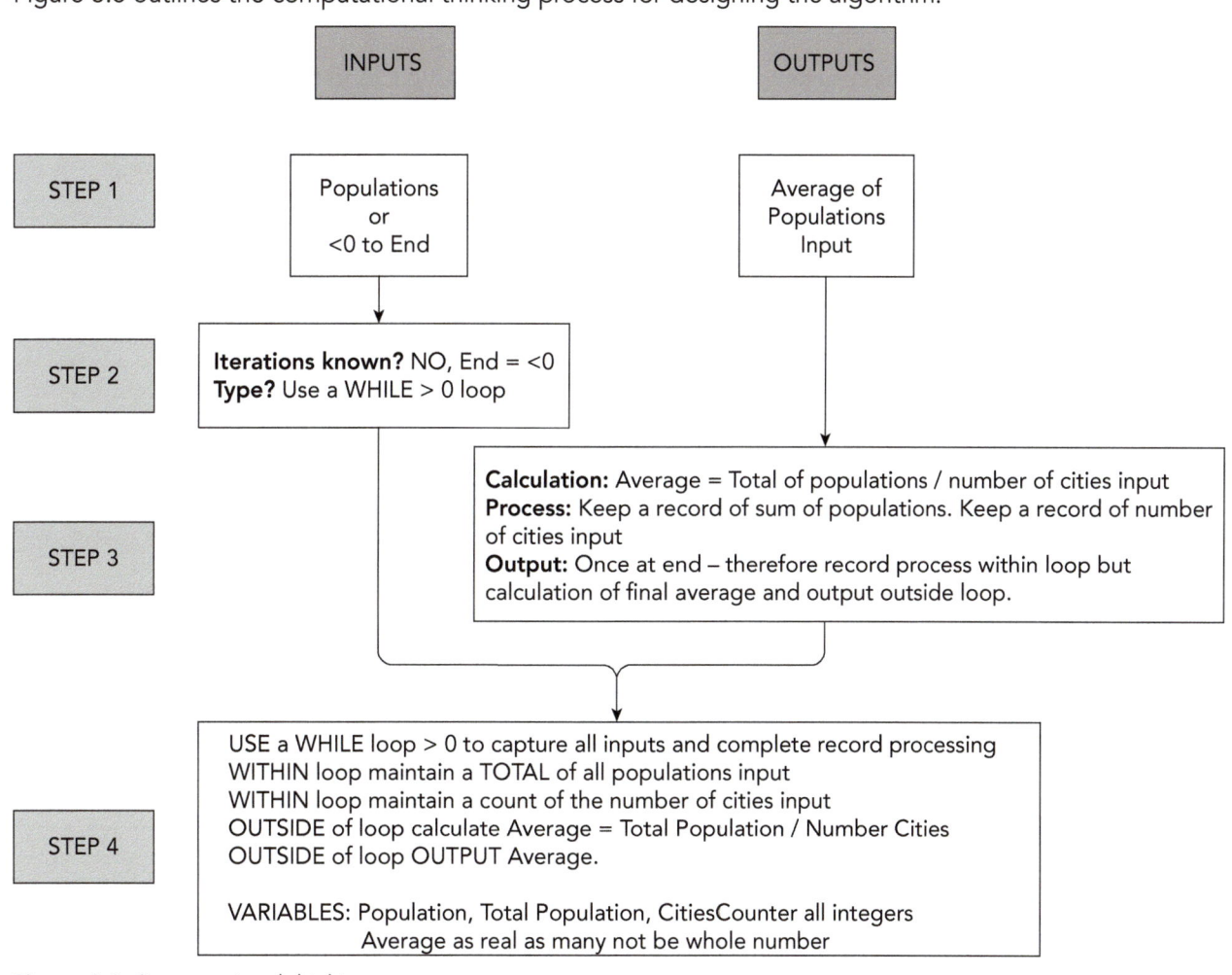

Figure 6.6: Computational thinking steps

CONTINUED

The structure diagram in Figure 6.7 shows a possible design for the system.

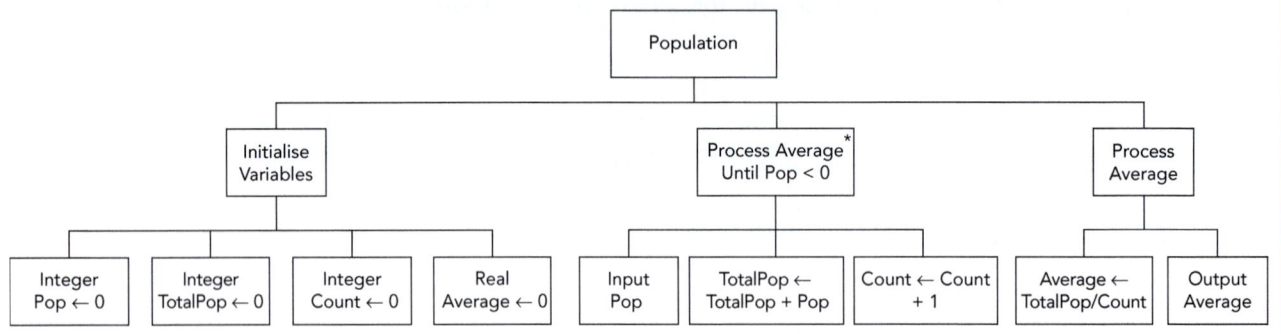

Figure 6.7: Structure diagram

From this, we can produce the resultant flowchart, shown in Flowchart 6.2, and the pseudocode.

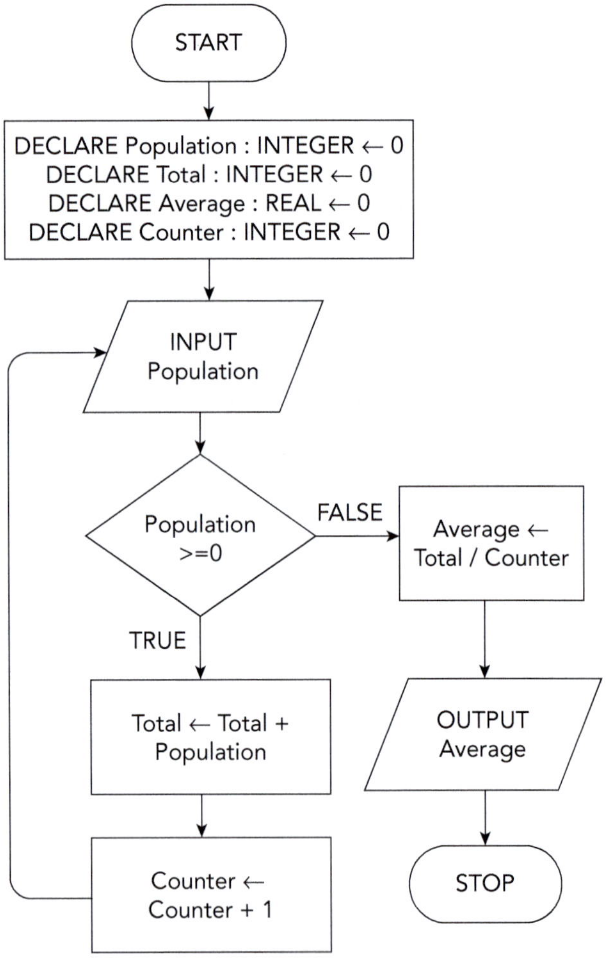

Pseudocode

```
DECLARE Population : INTEGER ← 0
DECLARE Total : INTEGER ← 0
DECLARE Average : REAL ← 0
DECLARE Counter : INTEGER ← 0

INPUT Population

WHILE Population >= 0 DO

    Total ← Total + Population
    Counter ← Counter + 1
    INPUT Population

ENDWHILE

Average ← Total/Counter
OUTPUT Average
```

Flowchart 6.2: Flowchart and Pseudocode

PRACTICE TASKS 6.3–6.4

6.3 Rainfall problem

A system is designed to collect monthly rainfall data in millimetres from weather stations around the country. It then has to output a monthly rainfall figure for each location and the average for the whole country. Draw a structured diagram that breaks this system down into subsystems. NOTE: There is more than one correct answer to this question.

6.4 Book system

A librarian has been given a set of books and is required to analyse the number of pages in the books in that set of books. The librarian will input the number of pages in each book, indicating the end of the input sequence by inputting a negative value. The system will output:

- the number of books that have fewer than 100 pages
- the number of books that have between 101 and 250 pages
- the number of books that have more than 250 pages
- the average number of pages in the set of books.

a Draw a structured diagram that breaks this system down into subsystems.

b Using pseudocode, produce an algorithm for a computerised solution to the task.

c Write and test a console application implementation of your algorithm.

> **TIP**
>
> To check that your solution works as intended, it should be tested. See Chapter 9 to discover how this can be achieved.

CHALLENGE TASK 6.1

Local rainfall

Local weather stations have been asked to enter the number of millimetres of rain falling in each day of the month when there is more than 0.1 mm of rain collected. There are therefore going to be a maximum of 31 entries and a minimum of 0 entries, but each station will enter a different number of data points. Design a program that will:

- input a list of daily rainfall totals (to the nearest 0.1 mm)
- end the input sequence when a negative number is entered
- output the total rainfall for the month.

Produce your algorithm as a flowchart or pseudocode and then write and test a console application implementation of your algorithm.

6.4 The complete design process

When presented with a complex problem, any or all of the following processes might be needed to create a finished solution:

- Decomposition – using top-down design and structured diagrams.
- Standard methods of solution – using flowcharts and pseudocode initially and then examining the algorithm to see if it could be simplified, or made more efficient by employing loops or subroutines.
- Check input data – This is called **validating** user input (see Chapter 8).
- Testing – thorough and effective **testing** needs to then take place (see Chapter 9).

Simplifying algorithms makes for more reliable and maintainable solutions. Efficient algorithms use loops and functions to collect repetitive input and processing. An efficient algorithm will combine the collection of multiple processes in one loop rather than in separate loops wherever possible.

You are expected to be able to design effective solutions and comment on the effectiveness of algorithms presented to you. Remember to check whether the solution could be simplified further or made more efficient. Check it handles all your test data (Chapter 9), and that all inputs are effectively validated (Chapter 8). The efficiency of different solutions was examined in more depth in Chapter 5 where the use of different kinds of loops was compared, and again in Chapter 12 where advice is given on how to prepare for programming scenario-based tasks.

SUMMARY

Computational thinking is the thought processes involved in expressing solutions as computational steps or algorithms that can be carried out by a computer. In effect, the essence of programming.
Systems are made up of subsystems, which may in turn be made up of further subsystems.
Top-down design is a method of simplifying the main system into its subsystems until the whole system is sufficiently defined to allow it to be understood and programmed.
Structure diagrams are a diagrammatical method of describing a system as a series of subsystems.
When designing solutions for a given problem, using an Input → Storage → Process → Output approach can help to design an effective solution.
Standard methods of solution are used when designing programmed systems. These are: flowcharts, pseudocode, top-down design, structure diagrams, simplification, efficiency analysis, validation and testing.
Effective solutions are efficient, have effective validation of inputs and have been tested to ensure they work as expected.

6 Designing algorithms

END-OF-CHAPTER TASKS

For each task, design an appropriate algorithm using a flowchart or pseudocode. Examples of working algorithms are included in the solutions, but remember these are not the only possible solutions.

1. A student is completing a mathematical probability study. They are required to throw a standard six-sided dice 100 times. They will input the number shown at each throw into a system. The system will output the number of times the dice shows a six and the average value of all the throws.

2. The student completing the mathematical probability study decides to extend the study to include two standard six-sided dice. The student will throw both dice simultaneously inputting both numbers shown at each throw into a system. The system will record the number of 'doubles' (throws that result in both dice showing the same number). The process will end when 100 doubles have been recorded, at which time the system will output the percentage of total throws that resulted in a double.

3. An experiment is undertaken to assess the reaction speed of 30 students. The time it takes for a student to react to an external stimulus is recorded. Each student has three attempts at the test. The reaction time of the student in all three tests is recorded. The average of the three reaction times is the final reaction time recorded for that student.

   The system is required to output the average of the final reaction times for the 30 students.

4. A manufacturer of canvas sunshades provide both rectangular and circular sunshades. Customers will often order more than one shade and the manufacture needs a system that will calculate the total cost of each order.

   The customer will be required to enter the type and dimensions of each sunshade they wish to order.
   - Rectangular sunshades – enter length and width in centimetres.
   - Circular sunshades enter the radius in centimetres.

   The customer will indicate the end of the order by entering the word 'END'. Once the input has been entered, the system will calculate the total cost of the order using the rules:
   - $5.25 for each square metre of canvas required – rounded up to the next full square metre
   - plus $10.00 for each shade made
   - plus a $20 delivery charge.

   The system will output the total cost of the order.

Chapter 7
Subroutines and file handling

IN THIS CHAPTER YOU WILL:

- learn how subroutines are used in programming
- understand how values are passed to and received from subroutines
- design, program and use a function
- design, program and use a procedure
- learn how a text file stores data
- understand how to store data in a text file
- understand how to read data from a text file.

Introduction

Any activity that is performed as an independent task within a larger process is a subroutine.

To cook a meal, you need to complete a number of smaller tasks:

1   You will have to buy the ingredients.
2   You may need to prepare some of the ingredients by peeling or soaking overnight.
3   You may have to start cooking the ingredients at different times, so they are all cooked by the meal time.
4   You will need to prepare the table and finally put the meal on plates or bowls.

Each of those processes is a subroutine used in the main task of solving the complex problem of cooking a meal to eat.

Routines are also used where a consistent approach is required. It is common for schools to teach students exactly how to act in a fire drill in the hope that, in the event of a real fire, students will act consistently and safely.

Computers make extensive use of subroutines in a similar way. They are used to provide coded solutions to specific tasks, often those commonly encountered. Those subroutines can be used by the main program code to maintain consistency and help complete the main task.

7.1 Why use subroutines?

A **subroutine** is a sequence of program code that performs a specific task but does not represent the entire system. The code can be activated by the main program code to complete specific tasks as and when needed.

When a subroutine is activated (this is known as '**called**'), the calling program is halted and control is transferred to the called subroutine. (NOTE: A subroutine can also call another subroutine.) After the called subroutine has completed execution, control is passed back to the calling program. This modularised approach to programming has a lot of advantages over a simple sequenced program.

- **The subroutine can be called when needed:** A single block of code can be used many times in the entire program, avoiding the need for repeating identical code sequences throughout the overall code. This improves the modularity of the code, makes it easier to understand and helps in the identification of errors.

- **The subroutine can be passed data to use in comparisons or calculations:** It can also return data to the main program or directly change data values held in the main program. This means a subroutine can be used to complete any task required.

- **There is only one section of code to debug:** If an error is located in a subroutine, only the individual subroutine needs to be debugged. Had the code been repeated through the main program, each occurrence would need to be altered.

- **There is only one section of code to update:** Improvements and extensions of the code are available everywhere the subroutine is called.

> **KEY WORDS**
>
> **subroutine:** subroutines provide an independent section of code that can be called from another routine while the program is running. In this way, subroutines can be used to perform common tasks within a program.
>
> **calling:** to activate a subroutine. To do this, you specify the subroutine's name and, optionally, parameters.

7.2 Subroutines in Visual Basic

Visual Basic makes extensive use of subroutines. You have already met IDE-generated subroutines.

When using console application, the system generates a main subroutine that will automatically run when the program is executed. All the code included within the subroutine is executed when the program is run.

```
Sub Main()
    'Code to execute
End Sub
```

Note: Windows Forms application is optional content. It is not covered in the syllabuses.

The Windows Forms application sets up individual subroutines for each event that is required, for example, a button click. The code within the event subroutine will be executed when the event is triggered.

```
Private Sub Button1_Click(sender As Object, e As EventArgs) Handles Button1.Click
    'Code to execute
End Sub
```

Although they look different, these subroutines have a number of common factors:

- They run the code contained within the subroutine when they are activated or called from another routine or the main program.
- They have a beginning indicated by the keyword `Sub` and an end indicated by the keyword `End Sub`.

It is possible to add subroutines to both console application and Windows Forms application.

Console application default is to include the one main subroutine `Sub Main()`, but as you will discover in this chapter, additional subroutines can be coded as required. One common approach is to use the main subroutine as a menu of options with each option directing the execution of the code to another subroutine.

The Windows Forms application provides inbuilt subroutines for a multitude of interface events. Using the drop-down selectors (see Figure 7.1) provides a list of the many event subroutines that are available.

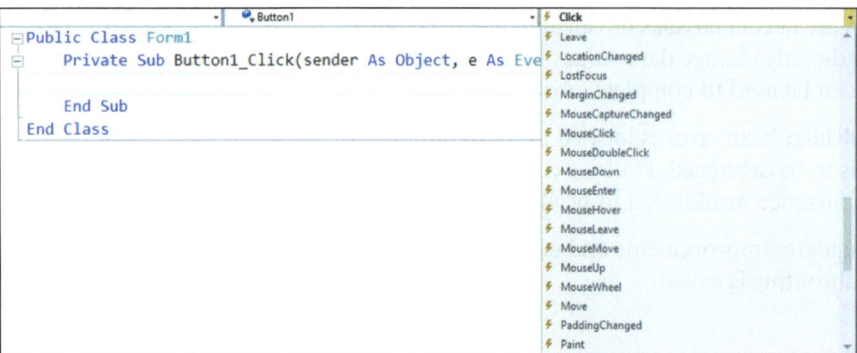

Figure 7.1: Event subroutine drop down list

There are two main types of subroutine: **functions** and **procedures**. Visual Basic supports both types of subroutine.

Although functions and procedures have common features, they also have unique individual features that make them suited to different tasks. The following sections discuss these features and identify how functions and procedures can be designed and included in your coded solutions to tasks.

7.3 Functions

A function can receive multiple data values but it will always return a single value.

The data values are passed using **parameters**, which are set up when the function is defined. A parameter is information about the data that the function is going to be using. Parameters can be either an actual data item or the identifier of a variable that holds the data item.

A function always returns a value through its identifier. The calling program will call a function and pass the required parameters. The function will complete execution and return a single value. This feature reduces the complexity of the code syntax, making a function the preferred subroutine for many tasks.

The format of a function will always include three elements. Code snippet 7.1 shows a simple function that takes two values as parameters and returns the result of multiplying those numbers.

```
FUNCTION Multiply (Num1 : INTEGER, Num2 : INTEGER) RETURNS REAL
    RETURN Num1 * Num2
ENDFUNCTION
```

Code snippet 7.1: The format of a function

Element 1

The declaration indicated by the keyword `FUNCTION`. This should define:

- The name of the function – `Multiply`.
- The parameter values to be passed to the function – `(Num1 : INTEGER, Num2 : INTEGER)`.
- The datatype of the value to be returned – `RETURNS REAL`.

Element 2

The `RETURN` keyword to indicate the value to be returned to the calling program.

Element 3

An indication of the end of the function – `ENDFUNCTION`.

The format for using a function is to assign the return value using the function identifier (name). This is similar to the way that a value is assigned to variable. The following pseudocode calls the function with the values 6 and 10 and stores the value returned by the function in a variable with the identifier '`Answer`'. The variable '`Answer`' would now hold the value 60.

```
Answer ← Multiply (6, 10)
```

> **KEY WORDS**
>
> **function:** a subroutine that can receive multiple parameters and returns a single value. A function always returns a value through its identifier.
>
> **procedure:** a subroutine that can receive and return multiple parameters. It may or may not return a value. If values are returned, they are returned via parameters.
>
> **parameter:** data or values that are passed to, or received from, a subroutine.

Although a function can only return one value to the calling program, it is possible for the function to have several return options. The function will use selection to determine which of the return options to follow. This can be seen in the pseudocode function 'OddOrEven', which will return the word 'ODD' or the word 'EVEN' following a calculation to identify if the number passed as a parameter is either an odd or even number.

```
FUNCTION OddOrEven (Number : INTEGER) RETURNS STRING
    IF Number MOD 2 = 0
      THEN
        RETURN "EVEN"
      ELSE
        RETURN "ODD"
    ENDIF
ENDFUNCTION
```

The code `OUTPUT OddOrEven(11)` would output the word 'ODD'.

Programming a function

DEMO TASK 7.1

Design a function that takes the radius of a circle as a parameter and returns the area of that circle.

Solution

Like all algorithms, it is good practice to design a function before programming. When designing a function, the standard flowchart symbols and pseudocode format are used.

Flowcharts show the logical flow of an algorithm but are not suited to showing the programming approach to achieving that logic. Consequently, they are not often used to define functions. Pseudocode is more suited to the role.

However, if a flowchart is used to define a function, it is common to have a separate flowchart for each function. The call to the function is indicated within the main code flowchart by using the subroutine or predefined process symbol. The function symbol is a rectangle with vertical bars either side. When this is met in a flowchart, it indicates that the design for the function will be independent. See 'Function Circle' in Flowchart 7.1.

7 Subroutines and file handling

> **CONTINUED**
>
> Pseudocode for main program:
>
> ```
> DECLARE Radius : REAL ← 0
> REAL Area ← 0
> OUTPUT "Enter Radius"
> INPUT Radius
> Area ← Circle (Radius)
> OUTPUT Area
> ```
>
> Pseudocode for function:
>
> ```
> FUNCTION Circle (Rad : REAL) RETURNS REAL
>     CONSTANT Pi ← 3.14159
>     RETURN Rad * Rad * Pi
> ENDFUNCTION
> ```
>
> [Flowchart: START → DECLARE Radius : REAL ← 0 → OUTPUT "Enter radius" / INPUT Radius → Function Circle → OUTPUT Circle → STOP]
>
> **Flowchart 7.1:** Flowchart for main program showing function call
>
> In this example, the datatype of the parameter has also been identified. `FUNCTION Circle (Rad: REAL) REAL` states that the datatype is REAL. While this is not mandatory, it is good practice.

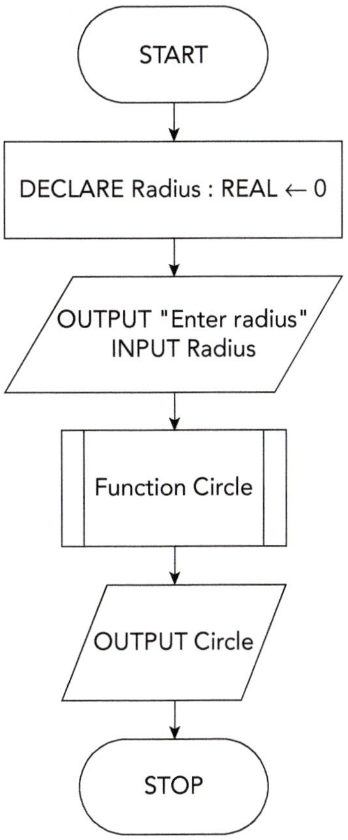

The Visual Basic syntax for a function

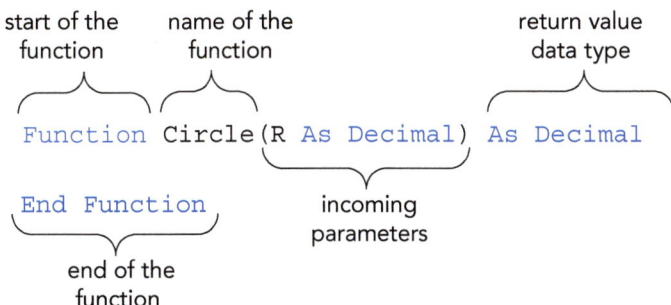

Figure 7.2: Visual Basic syntax for a function

The following is the console application code solution to Demo task 7.1. NOTE: A function cannot form part of another subroutine so the entire code must fall outside of any other subroutines.

```vb
Module Module1
    'The main code that will call the function
    Sub Main()
        Dim Radius As Decimal
        Dim Area As Decimal
        Console.WriteLine("Enter radius")
        Radius = Console.ReadLine()


        'The variable area is assigned the return value
        from the function
        'The function is passed the value within the
        variable 'Radius'
        Area = Circle(Radius)
        Console.WriteLine(Area)
        Console.ReadKey()
    End Sub


    'The function that takes a single decimal parameter
    Function Circle(Rad As Decimal) As Decimal
        Const Pi As Decimal = 3.14159
    Return Rad * Rad * Pi
    End Function
End Module
```

Code snippet 7.2: Circle function using parameters

It is also possible to use the returned value directly in the `Console.WriteLine` command.

```vb
    Sub Main()
        Dim Radius As Decimal
        Console.WriteLine("Enter radius")
        Radius = Console.ReadLine()
        Console.WriteLine(Circle(Radius))
        Console.ReadKey()
    End Sub
```

> **TIP**
>
> When multiplying or dividing numbers, it is possible that the result will contain several decimal places. You may notice this when you run the circle function. A decimal variable can represent a number that contains 29 digits with up to 28 decimal places, so it is not unusual that a programmer will wish to round the number before outputting. Visual Basic includes the `Math.Round` function to achieve this.
>
> The function takes two parameters, the number and the number of decimal places:
>
> `Math.Round(1.13291, 3)` will return 1.133 (1.13291 rounded to 3 decimal places).
>
> It is also possible to pass the function a calculation rather than an actual number. This could have been used effectively in the circle function.
>
> ```vb
>     Return Math.Round(Rad * Rad * Pi, 2)
> ```

7 Subroutines and file handling

Passing parameter values

A function can have any number of parameters. Some functions have no parameters at all and simply return a value. For example, the random number function used in Visual Basic `Rnd()` will return a random number between 0 and 1. The function still includes empty brackets to indicate that it is a function, but no parameter values are passed.

Another example where a function has no parameters is when any user input is obtained within the function. Consider this alternative to the 'Circle' function in Code snippet 7.2:

```vb
Module Module1
    Sub Main()
        'Call using the function name and empty brackets as
        no parameters
        Console.WriteLine(Circle())
        Console.ReadKey()

    End Sub


    'The function takes no parameter so empty brackets
    Function Circle() As Decimal
        Dim Rad As Decimal
        Const Pi As Decimal = 3.14159
        'User input needed by the function obtained within
        the function
        Console.WriteLine("Enter radius")
        Rad = Console.ReadLine()
        'The return value using Math.Round function to
        limit to 2 decimal places
        Return Math.Round(Rad * Rad * Pi, 2)
    End Function
End Module
```

Where parameters are to be used, the name of the parameter in the function and the name of the variable in the main code that holds the value to be used are unimportant. What is crucial is the order in which the parameters are passed. They will be received by the function in the order they are passed.

Consider this function:

```
FUNCTION Example (P1 : INTEGER, P2 : INTEGER, P3 :
  INTEGER) RETURNS INTEGER
    RETURN (P1 + P2) * P3
ENDFUNCTION
```

Should this function be called with direct values, the order in which they are passed will decide the parameter they represent. The function call `Example(2,3,4)` would result in a value of 20 being returned, because P1 = 2, P2 = 3 and P3 = 4.

The same applies when passing values held in variables. Even if the main code had variable names that matched the parameter names, it is the order in which the variables were passed that would determine the parameter values, not the matching names.

Using pseudocode, design the solutions to these practice tasks, before coding the system and testing it works as expected.

PRACTICE TASKS 7.1–7.3

7.1 Generate a function that takes as a parameter a single integer value and returns the square of that number. For example, if passed the value 4, the function would return 16. The function should be called from the main program subroutine.

7.2 Generate a function that takes two numbers as parameters and returns the higher number. The function should be called from the main program subroutine.

7.3 Generate a function that can check a new password meets certain criteria. It will take the password as a parameter value and return the word 'PASS' or the word 'FAIL', depending on whether the password meets the following criteria:

- Must have more than 6 characters.

> **TIP**
>
> Use the `.Length` command, which returns the number of characters in a string variable. For example, if a variable 'PW' contained the string value 'Password$1'. The command `PW.Length` would return the value 9.

- Must include the $ symbol.

> **TIP**
>
> Use the `.Contains()` command, which returns the Boolean value True if the defined character is in the string, or the Boolean value False if it is missing from the string. For example, if a variable 'PW' contained the string value 'Password$1'. The command `PW.Contains("$")` would return the value True.
>
> The function should be called from the main program subroutine.

CHALLENGE TASKS 7.1–7.2

7.1 Extend the password check (Practice Task 7.3) so that the user is required to make a choice. Either enter a password for checking or indicate they wish to exit the program. If they choose to check a password, the function will be called. If they choose to exit the system will close.

7.2 Generate a function that takes two numbers as parameters and returns the highest common factor of those numbers. For example, if passed 18 and 27 the function would return 9. (9 is the highest number that will divide exactly into both 18 and 27.)

> **TIP**
>
> Use a conditional loop in the main subroutine to allow multiple inputs.

7.4 Procedures

A procedure is a subroutine that can receive and return multiple data values. It is possible that a procedure does not return any value.

The data values are passed using parameters that are set up when the procedure is defined.

The calling program will provide any parameter values required by the procedure during its execution. The procedure may use the parameters to complete a task without returning any value to the calling program. Alternatively, once the procedure has been executed, it may make available any values resulting from its execution to the calling program. In this situation, the calling program will also pass references to the procedure to tell the procedure which variable or location it should copy the calculated values.

The main differences between procedures and functions are:

- Functions return a single value through their identifier.
- Procedures output zero or more values through their parameters.

The format of a procedure will always include two elements.

To show the format of a procedure, a comparison has been made with the Multiply function first used to explain the format of a function (Code snippet 7.1).

FUNCTION

```
FUNCTION Multiply (Num1 : REAL,
Num2 : REAL) RETURNS REAL
    RETURN Num1 * Num2
ENDFUNCTION
```

PROCEDURE

```
PROCEDURE Multiply (Num1 : REAL, Num2 :
REAL, Output : REAL)
    Output ← Num1 * Num2
ENDPROCEDURE
```

Similarities		

Start / End	Both functions and procedures have similar start and end statements.	`FUNCTION`
		`ENDFUNCTION`
		
		`PROCEDURE`
		`ENDPROCEDURE`

Naming	The name is included in the same place in both functions and procedures.	`FUNCTION` **`Multiply`**
		`ENDFUNCTION`
		
		`PROCEDURE` **`Multiply`**
		`ENDPROCEDURE`

Ongoing Parameters	Both functions and procedures define and receive the parameter in the same way.	`FUNCTION Multiply (Num1 : REAL,`
		`Num2 : REAL`
		
		`PROCEDURE Multiply (Num1 : REAL,`
		`Num2 : REAL`

(continued)

Differences		
Returning values	The function includes the value to be returned by the keyword RETURN.	`RETURN Num1 * Num2`
	The procedure does return values, It provides an outgoing parameter which can be used by the main program. The main program will pass the procedure the name of a variable that will receive the outgoing value.	`PROCEDURE Multiply (Num1 : REAL, Num2 : REAL, Output : REAL)` `    Output ← Num1 * Num2`

Table 7.1: Similarities and differences between functions and procedures

Programming a procedure

The syntax for setting up a procedure in Visual Basic is shown in Figure 7.3. As with functions, a procedure cannot form part of another subroutine, so the entire code must fall within the class or module but outside any other subroutines.

Visual Basic defines a procedure by the key word `Sub` (short for subroutine). It is usual in pseudocode to define with the keyword 'PROCEDURE'.

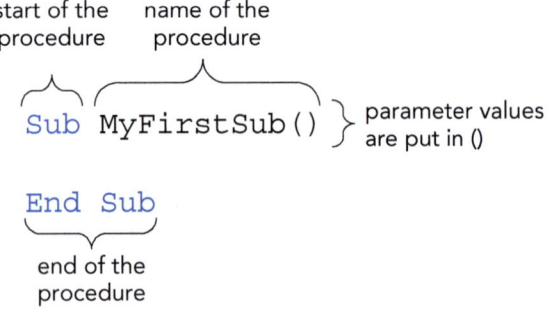

Figure 7.3: Visual Basic syntax for a procedure

A procedure is called from the main code as follows:

```
Sub Main()
    Call MyFirstSub()
End Sub
```

Passing parameters by value or reference

Functions pass their return value through their identifier. As a result, there is no need to consider whether parameters are ingoing (passing data to be used) or outgoing (references to the return location).

Procedures use parameters to identify both ingoing and outgoing values, so there is a need to be able to differentiate between ingoing and outgoing parameters.

Parameters can be **passed by value** or by **reference**.

KEY WORDS

passing by value: the subroutine holds a local copy of the data passed. Any changes made to the data are held in the local copy within the subroutine. If the data passed originated from a variable in the main code, the original value in that variable remains unaltered.

passing by reference: the subroutine is passed a value by reference to a variable or array declared in the main code that holds the data to be used. Any changes made to that value is stored in the referenced variable in the main code.

- If data is passed by value, a local copy of the data is held by the procedure. This local copy stores the result of any calculation completed within the procedure and is discarded when the procedure exits.
- If data is passed by reference, the location in memory – normally a variable identifier in the main program – that is passed to the procedure is used by the procedure. Changes due to calculations are stored directly in the variable and are maintained and used by the main program when the procedure exits.

In Visual Basic, the required method of passing parameters is indicated by the keywords described in Table 7.2.

Keyword	Use	Description
ByVal	ByVal Number As Integer	ByVal indicates the parameter is passed by value.
		Number is the identifier used within the procedure.
		The procedure will store a local copy of the data passed to the parameter in the variable Number. The variable Number cannot be accessed outside the procedure.
ByRef	ByRef Output As Integer	ByRef indicates the parameter is passed by reference.
		Output is the identifier used within the procedure to hold values. It will update any value in the variable reference passed from the calling program.

Table 7.2: Keywords used in passing parameters

Parameters that are intended to be output to the main program need to be passed by reference (prefixed by the keyword ByRef in Visual Basic). The order of the parameters is not important, but they need to be called in the same order they are defined. It is common practice for output parameters to be defined last in a procedure.

For example, consider the code below designed to calculate the average score achieved in a test:

```
Sub Div(TotalScore As Integer, NumStudents As Integer,
ByRef Average As Decimal)
    Average = TotalScore / NumStudents
End Sub
```

This would need to be called using the following main program:

```
Sub main()
    Dim TotalScore As Integer = 123
    Dim NumStudents As Integer = 12
    Dim Average As Decimal = 0
    Call Div(TotalScore, NumStudents, Average)
    Console.WriteLine(Average)
    Console.ReadKey()
End Sub
```

However, had the procedure been declared with the parameters in a different order – in this example, with the output parameter defined first – then the calling program would need to pass the receiving variable first.

```
Sub Div(ByRef Average As Decimal, TotalScore As Integer,
NumStudents As Integer)
    Average = TotalScore / NumStudents
End Sub


Call Div(Average, TotalScore, NumStudents)
```

Both these examples, if called correctly, would output the expected result. But it is crucial that the passing of parameters match the order they are defined in the procedure.

Example of the need for ByRef parameters

Consider the following code:

```
Sub Testing(A As Integer, B As Integer, C As Integer)
    C = A + B
End Sub
```

Code snippet 7.3: All parameters passed by value

> **TIP**
>
> If the `ByVal` or `ByRef` is omitted from either pseudocode or actual Visual Basic code, the default is that the parameter is passed by value. Where a value is to be returned, the inclusion of the `ByRef` is vital.

As the `ByRef` or `ByVal` has been omitted, the default of `ByVal` will be applied. This means that all the parameters values are held locally and no value is returned by the procedure.

Even though the local parameter 'C' is calculated within the procedure, as the parameter has been defined as `ByVal`, it cannot output a value to the main code.

The code should have been written as:

```
Sub Testing(A As Integer, B As Integer, ByRef C As Integer)
    C = A + B
End Sub
```

Code snippet 7.4: Parameter 'C' passed by reference

> **PRACTICE TASK 7.4**
>
> Reproduce the procedure using a console application. Call it from the main code (Code snippet 7.5), firstly with all parameters defined `ByVal` (Code snippet 7.3) and secondly with the parameter 'C' defined `ByRef` (Code snippet 7.4).
>
> What is the result of calling both versions of the procedure?
>
> ```
> Sub Main()
>     Dim N1, N2, N3 As Integer
>     Console.WriteLine("N1")
>     N1 = Console.ReadLine()
>     Console.WriteLine("N2")
>     N2 = Console.ReadLine()
>     Call Testing(N1, N2, N3)
>     Console.WriteLine(N3)
>     Console.ReadKey()
> End Sub
> ```
>
> Code snippet 7.5: Example code for calling a procedure

7 Subroutines and file handling

DEMO TASK 7.2

A procedure is required that will take two integer values as parameters. It will output the quotient and modulus results of integer division of the first number by the second number.

Solution

MAIN PROGRAM

```
DECLARE Num1 : INTEGER ← 0
DECLARE Num2 : INTEGER ← 0
DECLARE Quotient : INTEGER ← 0
DECLARE Modulus : INTEGER ← 0
Call QuoMod (Num1, Num2, Quotient, Modulus)
OUTPUT Quotient
OUTPUT Modulus
```

PROCEDURE

```
PROCEDURE QuoMod(N1 : INTEGER, N2 : INTEGER, ByRef Q :
  INTEGER, ByRef M : INTEGER)
    Q ← N1 DIV N2
    M ← N1 MOD N2
ENDPROCEDURE
```

NOTE: (in the procedure) the passing of the two outgoing parameters as `ByRef`.

```vb
Module Module1
    Sub Main()
        Dim Num1 As Integer = 0
        Dim Num2 As Integer = 0
        Dim Quotient As Integer = 0
        Dim Modulus As Integer = 0

        Console.WriteLine("Enter first number")
        Num1 = Console.ReadLine()
        Console.WriteLine("Enter second number")
        Num2 = Console.ReadLine()
        Call QuoMod(Num1, Num2, Quotient, Modulus)
        Console.WriteLine(Quotient)
        Console.WriteLine(Modulus)
        Console.ReadKey()
    End Sub

    Sub QuoMod(ByVal N1 As Integer, N2 As Integer,
    ByRef Q As Integer, ByRef M As Integer)
        Q = N1 \ N2
        M = N1 Mod N2
    End Sub
End Module
```

> **CONTINUED**
>
> NOTE how the definition of the parameter 'N1' and 'N2' differ. The first parameter 'N1' has been defined using the full `ByVal` definition. The second parameter 'N2' is also a `ByVal` parameter but, as no specific definition has been made, the default value of `ByVal` has been applied.
>
> An alternative method of achieving the same output as the procedure `QuoMod` shown above would be for the procedure to include all the code: the code to obtain the required input, the code to calculate the required values and the code to provide the output to the user.
>
> ```
> Module Module1
>     Sub Main()
>         Call QuoMod()
>         Console.ReadKey()
>     End Sub
>
>
>     Sub QuoMod()
>         Dim N1 As Integer = 0
>         Dim N2 As Integer = 0
>         Dim Quotient As Integer = 0
>         Dim Modulus As Integer = 0
>         Console.WriteLine("Enter first number")
>         N1 = Console.ReadLine()
>         Console.WriteLine("Enter second number")
>         N2 = Console.ReadLine()
>         Quotient = N1 \ N2
>         Modulus = N1 Mod N2
>         Console.WriteLine(Quotient)
>         Console.WriteLine(Modulus)
>     End Sub
> End Module
> ```
>
> NOTE: The procedure includes empty brackets to indicate that it is a procedure without parameters. When the procedure is called in the main code, it is called using the parameter name with the empty brackets.

Using pseudocode, design the solutions to these tasks before coding the system and testing it works as expected. (Practice Tasks 7.5 and 7.6 are a repeat of the function tasks. Compare your code with the code you generated for the function tasks to remind yourself how functions and procedures differ.)

> **PRACTICE TASKS 7.5–7.7**
>
> **7.5** Generate a procedure that takes as a parameter a single integer value and outputs the square of that number. For example, if passed the value 4 the procedure would return 16. The procedure should be called from the main program subroutine.
>
> **7.6** Generate a procedure that takes two numbers as parameters and outputs the higher number. The procedure should be called from the main program subroutine.
>
> **7.7a** Generate a procedure that takes as a parameter the radius of a circle. The procedure will output both the circumference and the area of the circle. The procedure should be called from the main program subroutine.
>
> **b** Alter this so that all the input and output code is included within the procedure. The procedure should be called from the main program subroutine.

7.5 File handling

So far in this book we have used variables to hold data values while a program is running. This will work effectively while a system is running but once the system is exited the data will be lost. The next time the program is run the variables will hold the initialisation value that had been coded. Imagine a computer game that could not store players progress when they exited the game. Each time they played the game they would have to start from the beginning.

One way of storing data permanently is to use external files. Data from a program can be written (saved) to a file and the program can then be exited. This storage and access of data is called **file handling**. When the program is run again the data can then be read from the file and stored back in the variables.

Visual Basic has a library of code that will support you in writing to and reading from files.

Text files

It is possible with Visual Basic to save data to many different types of external files and databases. In this book we will use a basic text file to hold our data.

A **text file** is a simple form of data storage. It stores the data in a text format and is normally saved with the .txt extension.

Writing data to a file

The process of **writing data to a file** involves the following steps:

1. Open the file so it can have data written to it.
2. Write the data to the file.
3. Close the file once the writing process has been completed.

> **KEY WORDS**
>
> **file handling:** programming statements that allow text files to be opened, read from, written to and closed.
>
> **text file:** a file that stores data as text. It normally has the file extension .txt.
>
> **writing data to a file:** the process of saving data to a text file.

The way in which different languages complete this process can be involved but fortunately the pseudocode for the process is reasonably simple.

Table 7.3 shows the pseudocode to write the data from a variable `HighScore` to a file `Score.txt`.

Process step	Pseudocode
Open the file	`OPENFILE Score.txt FOR WRITE` The file is opened for writing. It is important that a file is not just opened, the mode of operation (write or read) must be included.
Write data to the file	`WRITEFILE Score.txt, HighScore` The file that is to be written to should be included. The variable that holds the data that is to be written is defined after the name of the file, separated by a comma.
Close the file	`CLOSEFILE Score.txt` The file is closed. If this step is omitted the file is locked and cannot be accessed by another part of the program. In the real programming code if this step is omitted the write process will not be completed. When the program is exited the file will be closed without the data having been saved.

Complete pseudocode	`OPENFILE Score.txt FOR WRITE`
	`WRITEFILE Score.txt, HighScore`
	`CLOSEFILE Score.txt`

Table 7.3: Pseudocode to write the data from a variable

Writing to a file in Visual Basic

Visual Basic contains a library of code to help you write to a text file, **StreamWriter**. The code is contained in the `System.IO.StreamWriter` library. To access the code, you will need to declare a variable that will act as your local version of the library.

The code to set up the variable would look like:

`Dim MyWriter As New System.IO.StreamWriter("C:\Example.txt")`

Each part of this line of code is important – Table 7.4 explains each part.

> **KEY WORD**
>
> **StreamWriter:** a class library within Visual Basic that provides access to code that is used to create text files and save data to text files.

Code	Explanation
`Dim MyWriter`	`MyWriter` is the name of the local variable you will use to access and write to the file.
`As New System.IO.StreamWriter`	This is different to a normal variable which is given a data type. It makes use of object orientation and classes which is beyond the scope of this book. The `As New` indicates that the variable is to become a local object of the `StreamWriter` class. In simple terms this gives the variable `MyWriter` the ability to use the code contained in the `SteamWriter` library.
`("C:\Example.txt")`	The declaration requires at least one parameter. `"C:\Example.txt"` is the file name of the file to be opened, this must be included. Note that the file name is contained in quotes. If a file path is not shown the system will save the file in the 'bin' folder of the program.

Table 7.4: Library of code

Now that the local object is set up it can be used to write to and close the file. The individual `writeline` and `close` commands are accessed through the use of the . as is common in other Visual Basic libraries.

The following code shows a full write and close command using the text contained in a variable called `DataToWrite`. The file to be used is called `Example` and is stored in the Visual Basic program as no file path has been given. The file is opened in overwrite mode so any existing data will be overwritten by the write command.

```
Dim DataToWrite As String = "Hello"

Dim MyWriter As New System.IO.StreamWriter("Example.txt")
MyWriter.WriteLine(DataToWrite)
MyWriter.Close()
```

To view a file saved within the bin folder you will have to select the 'Show All Files' option and then navigate to the bin / Debug folder (Figure 7.4). Selecting the file will display the contents of the file as a new tab in the main window.

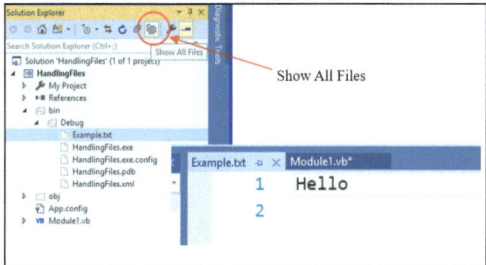

Figure 7.4: Viewing a saved file

Reading data from a file

The process for **reading data from a file** is similar to that for writing data to a file.

Table 7.5 shows the pseudocode to read data from a file called `Score.txt` and to save it in a variable called `StartHighScore`.

> **KEY WORD**
>
> **reading data from a file:** the process of reading the data held in the text file.

Process step	Pseudocode
Open the file	`OPENFILE Score.txt FOR READ` The file is opened for reading. It is important that a file is not just opened, the mode of operation (write or read) must be included. The only difference between this and the write code is the mode of operation indicated.
Read data from the file	`READFILE Score.txt, StartHighScore` The file that is to be read from should be included. The variable that is to hold the data read from the file is defined after the name of the file, separated by a comma. The only difference between this and the write command is the word `READFILE`. The order of the file and variable in the statement are the same in both processes.

(continued)

Process step	Pseudocode
Close the file	`CLOSEFILE Score.txt` The file is closed. This is identical to the write pseudocode.

Complete pseudocode	`OPENFILE Score.txt FOR READ`
	`READFILE Score.txt, StartHighScore`
	`CLOSEFILE Score.txt`

Table 7.5: Pseudocode to read data from a file

Reading from a file in Visual Basic

Visual Basic contains a library of code to help you read from a file, StreamReader. To read from a file we make use of the code contained in the `System.IO.StreamReader` library. The code follows a similar approach to that used to write to a file. The following code shows the full read process and the resultant output in the console window in Figure 7.5.

```vb
Dim ReceivingVariable As String = ""

Dim MyReader As New System.IO.StreamReader("Example.txt")

ReceivingVariable = MyReader.ReadLine()
MyReader.Close()

Console.WriteLine("This should be the data in the file")
Console.WriteLine(ReceivingVariable)
```

```
This should be the data in the file
Hello
```

Figure 7.5: Console output

> **PRACTICE TASK 7.8**
>
> **a** This task is to produce a program in pseudocode that will allow a user to input and save their first name to a text file called `MyName.txt`.
> Write a console application version of your program. You may save the file at any location you wish. Remember if you do not want the file to save in the same location as the program you will need to give the full file path.
>
> **b** Extend both your pseudocode and programming code so that the data saved in the file `MyName.txt` can be read and output to the user.

Validating the existence of a data file

If you try to read data from a file that does not exist, you will find that your program will output an error message (Figure 7.6).

Exception Unhandled

System.IO.FileNotFoundException: 'Could not find file

Figure 7.6: Error message

To avoid this error, it is possible to check the file exists before attempting the read command. The following program code shows how this can be achieved using the `Exists()` function in the `System.IO.File` library.

```
'Use the Exists function in the System.IO.File library
    'The condition in the IF statement checks if the return
    'from this is FALSE.
    If System.IO.File.Exists("Example3.txt") = False Then
        'If the return is FALSE a message is output
        Console.WriteLine("The file does not exist")
    Else
        'The ELSE path will only be followed if the the
        'return from the function is TRUE
        Dim ReceivingVariable As String = ""
        Dim MyReader As New System.IO.StreamReader("Example3.txt")
        ReceivingVariable = MyReader.ReadLine()
        MyReader.Close()
        Console.WriteLine(ReceivingVariable)
    End If
```

SUMMARY

Subroutines provide an independent section of code that be called from another routine while the program is running. In this way subroutines can be used to perform common tasks within a program.

As an independent section of code, a subroutine is easier to debug, maintain or update than repetitive code within the main program.

Subroutines are called from another routine. Once they have completed execution, they pass control back to the calling routine.

A function is a type of subroutine which can receive multiple parameters. It always returns a single value through its identifier.

A procedure is capable of receiving and returning multiple parameters. It may or may not return a value. If values are returned they are returned via a parameter.

The main differences between procedures and functions are:

- Functions return a single value through their identifier.
- Procedures return zero or more values through their parameters.

Subroutines can be passed values known as parameters. They can also pass parameters back to the calling program.

Parameters can be passed by value (`ByVal`) or by reference (`ByRef`). By value parameters hold a local copy of the value passed. By reference parameters are passed the reference to an external data location that they use as a variable. It is crucial that return parameters in a procedure as defined as `ByRef`.

Text files provide a method of permanently storing data. They store data as text and usually have .txt extension.

Before a file can be used, it must be opened.

When opening a text file, it is crucial that the required method of operation (write or read) is indicated.

Once a read or write operation is complete, the text file must be closed.

A loop structure can be used to write multiple values to a text file. This can be helpful when permanently saving data from an array.

END-OF-CHAPTER TASKS

Select either a function or procedure to complete the following tasks. Design your solution using pseudocode before coding and testing your design using Visual Basic.

1 A subroutine is needed that can be called from the main program. The subroutine will convert a time measured in seconds to show the same time in minutes and seconds. The program will pass the subroutine a single integer value that represents the number of seconds. The subroutine will return the appropriate minutes and seconds.

   For example, if passed 190 seconds the subroutine would return 3 minutes 10 seconds.

CONTINUED

2. A subroutine is needed that can be called from a main program. The subroutine will convert measurements of speed in kilometres per hour (kph) to miles per hour (mph). The main program will pass the subroutine the speed value in kilometres per hour and the subroutine will return the same speed in mile(s) per hour.

> **TIP**
> 1.60934 kilometres per hour = 1 mile per hour.

3. A subroutine is needed that can be called from the main program. The main program will pass the subroutine three different positive numbers. The subroutine will first check the numbers meet the criteria and return the value −1 (negative 1) if the numbers are not different and positive. If the numbers pass the check the subroutine will return the highest of the numbers input.

4. A subroutine is needed that can be called from the main routine. The main program will pass the subroutine a single integer value. The subroutine will display directly to the screen all the factors of that integer value.

    A factor of a number is an integer value that will divide exactly into the original number.

    For example, the factors of 18 are 1, 2, 3, 6, 9 and 18.

5. A system is required to calculate the amount of paint that is required to paint a wall. The main program will make use of a subroutine to complete the task.

    The subroutine will be called by the main program and passed the height and width of the wall.

    The subroutine will be required to calculate:

    a The exact amount of paint required using the ratio of 1 litre of paint to 8 square metres of wall.

    b The number of tins of paint needed to paint the wall. Paint can only be obtained in 5 litre tins.

    The subroutine will return two values:

    a The exact amount of paint required in litres.

    b The number of tins required to paint the wall.

6. A program is required that will prompt a user to input 10 numbers. Once the input has been completed:
    - The largest number input is to be saved to a file called `Large.txt`.
    - The average of the values input is to be saved to a file called `Ave.txt`.

    a Prepare an appropriate algorithm for this task in pseudocode.

    b Program and test a console application version of your algorithm.

7. Prepare a program in pseudocode that can read a value from a text file called `File1.txt` and then save that value in another text file called `File2.txt`.

8. A program is required that will allow a user to input a word. If the word input is longer than the word currently held in the file `LongWord.txt` then the value in the file is overwritten with the new word. The program should output an appropriate message to tell the user what has happened.

    Prepare an appropriate program in pseudocode for this task.

Chapter 8
Checking inputs

IN THIS CHAPTER YOU WILL:

- understand the need for accuracy of inputs
- know how to design validation routines using flowcharts or pseudocode
- understand the role and use of a range of validation and verification techniques:
  - presence check
  - range check
  - length check
  - type check
  - format check
  - check digit
- know how to program validation into your algorithms.

Introduction

Organisations rely on the accuracy of their data when making decisions. Inaccurate data can result in poor decisions, possibly with devastating results. Consider the situation of a doctor receiving inaccurate medical data about a patient, or a firefighter being given inaccurate data about wind speed and direction. The largest source of inaccuracies is during the data entry process. Therefore, it is important that systems are designed to help increase the accuracy of data entry.

8.1 Validation

Validation is the process of programming a system to automatically check that data falls within a set of specified criteria. Whilst validation cannot guarantee that data entered is accurate, it does ensure that it is reasonable. Systems should also filter out obvious mistakes. For example, if a system were recording the height of students, it would be reasonable to expect that they were all under 3 metres tall. Programming the system to reject data entries above 3 metres would help to remove obvious errors. However if a student's height was measured at 1.4 metres, but inaccurately entered as 1.04 metres, the system would still accept the value because it meets the validation criteria. Table 8.1 shows different types of validation checks:

> **KEY WORD**
>
> **validation:** the process of programming a system to automatically check that data satisfies a set of specified input criteria; for example, passwords must be longer than six characters.

> **TIP**
>
> Validation does not make data input accurate – this is a common misconception.

Validation type	Description	Examples
Presence check	Checks that required data has been input. The system will reject groups of data where required fields have been left blank. This is often used with data-collection forms.	Online order where 'Email Address' must be provided.
Range check	Checks data falls within a reasonable range. Data outside the expected range is rejected. It is possible to have data where the range limit is only applicable to one extreme. For example, the volume of a vessel cannot be zero but may not have an upper limit. This is known as a 'limit check'.	Age must be between 0 and 130 inclusive. Day of the month must be between 1 and 31 inclusive. Percentage score in an exam must be between 0 and 100 inclusive.
Length check	Checks that data entered is of a reasonable length. Data items that have a length outside the expected values are rejected. Normally used with text-based inputs.	Surname must be between 1 and 25, inclusive, characters long. A password must have more than six characters.
Type check	Checks that a data item is of a particular data type. It will reject any input that is of a different type.	Stock items in a shop must be entered as an Integer. Age will be numeric (e.g. it will not accept 'over 21').

(continued)

Validation type	Description	Examples
Format check	Checks that a data item matches a predetermined pattern and that the individual elements of the data item have particular values or data types.	Date of birth will be in the format dd/mm/yyyy. Mobile telephone numbers will be in the format NNNNN NNNNNN where N is a digit.
Check digit	Checks that a numerical data item has been entered accurately. Extra digit(s) are added to the number based on a calculation that can be repeated. This enables the number to be checked by repeating the calculation and comparing the calculated check digit with the value entered.	A barcode includes a check digit. ISBNs (book numbers) include a check digit.

Table 8.1: Validation checks

8.2 Verification

Verification is a process that helps to confirm the integrity of data as it is input into the system or when it is transferred between different parts of a system. **Data integrity** refers to the correctness of data during and after processing. While the format of the data may be changed by processing, if data integrity has been maintained, the data will remain a true and accurate representation of the original. Copying data should clearly not change the data values.

Two common forms of verification are visual check and double entry.

Visual check is a manual process completed by the person inputting the data. It is a check that is completed once data has been entered into the system. It involves checking that the data input into the system matches the original data. For example, if the data had been transferred into the system from a paper form. The data in the system would be visually checked against the original data on the paper form.

In double entry verification the data item is entered twice, often by different operators. The system will automatically compare the two input values to ensure they match. Double entry is a common feature encountered when setting a password, with the system requiring the user to enter the same password twice before accepting the new password.

The pseudocode for double entry verification could involve a simple IF statement within a REPEAT...UNTIL loop

```
REPEAT
    OUTPUT "Please enter your password"
    INPUT Password1
    OUTPUT "Please re-enter the password
    INPUT Password2
    IF Password1 <> Password2
      THEN
        OUTPUT "Passwords do not match - try again"
    ENDIF
UNTIL Password1 = Password2
```

> **KEY WORDS**
>
> **verification:** a process that confirms the integrity of data as it is input into the system or when it is transferred between different parts of a system; for example, a CAPTCHA image used to prove data is being entered by a human.
>
> **data integrity:** the correctness of data during and after processing.

It is a common mistake to believe that verification will make data accurate. While verification will help it cannot guarantee the accuracy of data. The visual check may miss errors or the original data may have been inaccurate and those inaccuracies have simply been copied in to the system. While double entry will ensure the two data entries match it could be that the same error had been made twice. For example, if the caps lock button on the keyboard had been selected by mistake for both passwords input, they would match but would not necessarily be the password expected.

8.3 Programming validation into your systems

Running code without passing the correct values to the variables will cause a program to crash or provide unexpected results. Run any of your programs without inputting the required numeric values, and you will receive the error message shown in Figure 8.1.

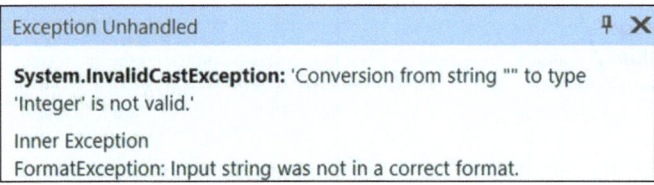

Figure 8.1: Invalid data type error message

As no value was input, the system has taken the empty input as an empty string value. It is not possible to convert an empty string to a numeric value and so this causes an unhandled exception error. Should this happen in a published program, the system would crash unexpectedly.

To avoid this type of error, validation could be used to check the data item before processing.

Presence check validation

> **DEMO TASK 8.1**
>
> *Code is required to check whether a user has input a value into a required field (firstname) in an online form. If the data is present, it will pass the check. NOTE: There is no guarantee that the name supplied will be in the correct format, so additional validation may be required.*

CONTINUED

Solution

Our algorithm for this task needs to start with inputting the user's first name. The validation process can then take place in a loop. The idea is that, while the data input is incorrect, the algorithm needs to keep giving the user another chance to enter the data correctly. When the test is passed, the program can continue. In this case, the test is simply whether there is any data. The algorithm can be seen in Flowchart 8.1.

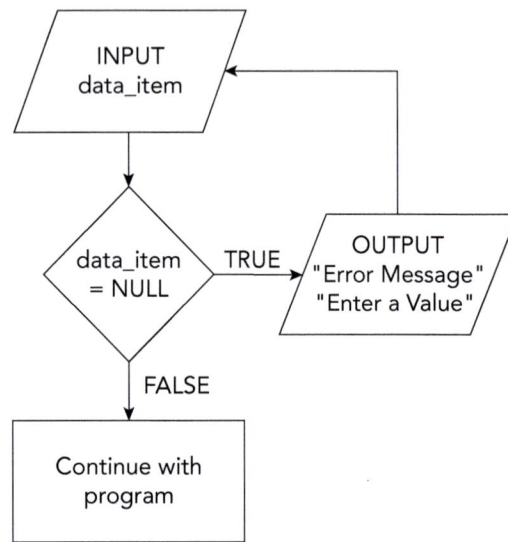

Flowchart 8.1: Presence check flowchart

In text-based programs, like the console application of Visual Basic, a WHILE loop can be used to check the presence of the input. If the data is missing, then the user can be prompted again to enter the required data. This requires the `Console.ReadLine()` method to pass the data before the check can be made.

```
Dim TextValue As String = ""
Console.WriteLine("Insert text value")
TextValue = Console.ReadLine()
Do While TextValue = ""
    Console.WriteLine("Insert text value")
    TextValue = Console.ReadLine()
Loop
```

> **CONTINUED**
>
> This method will work well with String data types, but it causes numeric data types to give an error – as shown in Figure 8.1. One solution is to make use of a temporary `String` variable to check for the presence of the data, before it passes the data onto the numeric variable, as shown below:
>
> ```vb
> Dim TextValue As String = ""
> Dim IntValue As Integer = 0
> Console.WriteLine("Insert text value")
> TextValue = Console.ReadLine()
> Do While TextValue = ""
>     Console.WriteLine("Insert text value")
>     TextValue = Console.ReadLine()
> Loop
> IntValue = TextValue
> ```
>
> **Note:** Windows Forms application is optional content. It is not covered in the syllabuses.
>
> In Windows Forms applications, the execution of code is triggered by an event such as a button press. This trigger cannot be controlled from within a WHILE loop in the same way as `Console.ReadLine()` can be controlled in console application. A different approach is therefore needed.
>
> In Windows Forms applications, the user is expected to input data prior to triggering the event. It is possible to check if the user input on the GUI has been completed. If the user input is missing, the program can output an error message and end the subroutine. This forces the user to input data before retriggering the event (i.e. pressing the button). If data is entered, that data is passed to the variables.
>
> The following code uses an IF statement. If data is entered, the IF statement executes the ELSE code, and the data is passed to the variables. If data is not entered, the IF statement runs the THEN code, outputs an error message and ends the subroutine.

CONTINUED

```vb
Private Sub BTNInput_Click(sender As Object,
e As EventArgs) Handles BTNInput.Click
        Dim InputText As String = ""
        If TBInput.Text = "" Then
            MsgBox("An input is required")
            'The Exit Sub command stops the execution of
            the subroutine, allowing the
            'user to correct errors before re-running the
            subroutine via the Button
            Exit Sub
        Else
            InputText = TBInput.Text
        End If
    End Sub
```

TIP

`Exit Sub` is a useful command in Visual Basic. When executed, it will stop the execution of whichever subroutine is running. This is a useful alternative to a complex conditional loop.

PRACTICE TASKS 8.1–8.2

8.1 Last name

a Produce a flowchart and the pseudocode for a system that will ask for a last name, and provide an error message if it is not present.

b Write a console application implementation of your algorithm.

8.2 Presence checks

Go back to some of the code you have generated for previous chapters. Include presence checks for the inputs in those systems.

CHALLENGE TASK 8.1

Checking a form

A Windows Forms application takes the first name, last name and date of birth of a user. The system is required to complete a presence check on each input element and output a specific message to inform the user which input elements are empty.

Produce the flowchart and pseudocode for a system. Write and test a Windows Forms application solution for your algorithm.

Range check validation

DEMO TASK 8.2

A system is required that asks the user to input the day of the month. The system should only accept numbers from 1 to 31 and reject all others.

CONTINUED

Solution

The principle for this algorithm is the same as for the presence check validation technique. Our algorithm needs to keep offering another chance for the user to enter their input while it does not pass our test. In this case, it is a simple check to see if the data is less than 1 or more than 31. If it is, then the algorithm needs to loop back to the request for user input. When the data is in range, the program can continue. This is illustrated in Flowchart 8.2 with the pseudocode alongside.

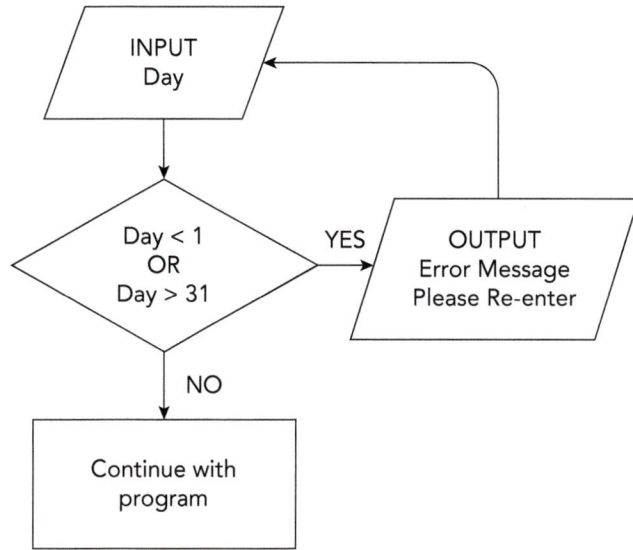

Here is the pseudocode solution (note, again, the use of a WHILE loop):

```
INPUT Day
WHILE Day < 1 OR Day > 31 DO
    OUTPUT Error Message
    INPUT Day
ENDWHILE

Continue with program
```

Flowchart 8.2: Flowchart for range check

The algorithm in Flowchart 8.2 (and the pseudocode) can be programmed in console application like this:

```
Sub Main()
    Dim Day As Integer = 0
    Console.WriteLine("Insert Day")
    Day = Console.ReadLine()
    Do While Day < 1 Or Day > 31
        Console.WriteLine("Insert Day between 1 and 31")
        Day = Console.ReadLine()
    Loop
    'program continues
End Sub
```

TIP

WHILE loops check criteria before running. The criteria are defined to identify inputs outside the required range. If an acceptable value is input, the loop never runs and the program continues. If an input is outside the expected range, the loop continues to iterate. This effectively halts the program until an acceptable value is input.

> **PRACTICE TASKS 8.3–8.4**
>
> **8.3 What month?**
>
> a Produce a flowchart and the pseudocode for a system that will ask for the month and provide an error message if it is not between 1 and 12 inclusive.
>
> b Write and test a console application implementation of your algorithm.
>
> **8.4 Day of the week**
>
> a Produce a flowchart and the pseudocode for a system that will ask for the day of the week and provide an error message if it does not match one of: Monday, Tuesday, Wednesday, Thursday, Friday, Saturday or Sunday.
>
> b Write a console application implementation of your algorithm.

Length check validation

> **DEMO TASK 8.3**
>
> A system is required that asks the user to input a password of six or more characters. The system should only accept passwords with six or more characters. It should produce an error message if the submitted password contains five characters or fewer.
>
> **Solution**
>
> The code needs to calculate the length of the password. It then follows a similar process to a range check using a WHILE loop to check the input against the required criteria. If the data input is acceptable, the system should continue to run; if not, the system will output an error message to the user. Flowchart 8.3 shows the flowchart design for the algorithm with the pseudocode alongside.
>
>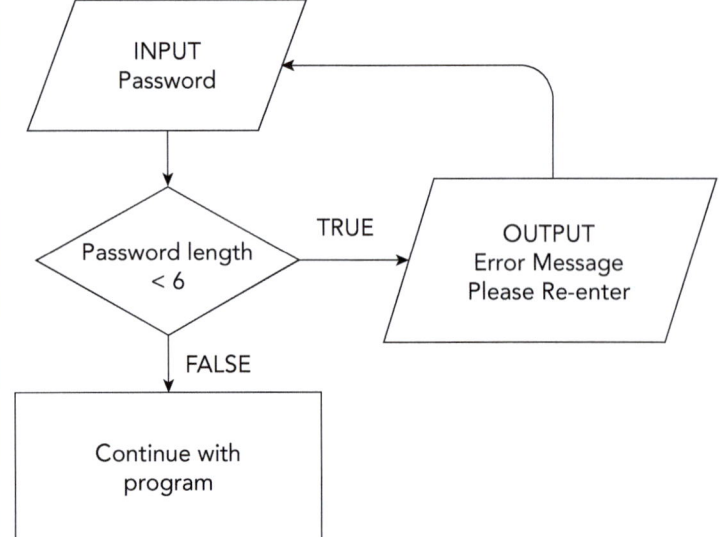
>
> Here is the pseudocode solution:
>
> ```
> DECLARE Password : STRING ← ""
> OUTPUT "Enter Password"
> INPUT Password
>
> WHILE LengthofPassword < 6 DO
>     OUTPUT "Password must be
>     6 characters"
>     INPUT Password
> ENDWHILE
>
> Continue with program
> ```
>
> Flowchart 8.3: Flowchart for length check

> **CONTINUED**
>
> The algorithm in Flowchart 8.3 (and the pseudocode) can be programmed in console application like this:
>
> ```vb
> Sub Main()
>     Dim Password As String = ""
>     Console.WriteLine("Enter Password")
>     Password = Console.ReadLine()
>     'The .length will return the number of characters in a string
>     Do While Password.Length < 6
>         Console.WriteLine("Password must be 6 characters")
>         Password = Console.ReadLine()
>     Loop
>     'program continues
> End Sub
> ```

> **TIP**
>
> The `.length` feature returns the number of characters in the string. It only works with string values, not any other data type.

> **PRACTICE TASKS 8.5–8.6**
>
> **8.5 Mobile phone number length**
>
> a   Produce a flowchart and the pseudocode for a system that will ask for a mobile phone number. If the mobile phone number does not have the correct number of digits, an error message is shown and the user is prompted to re-enter the mobile phone number.
>
> b   Write a console application implementation of your algorithm.
>
> **8.6 Maximum and minimum username length**
>
> a   Produce a flowchart and the pseudocode for a system that will ask for a username. The username must be longer than 5 characters and fewer than 11 characters. If the username input does not meet this criteria, an error message is shown and the user prompted to re-enter the username
>
> b   This task combines which two forms of validation?
>
> c   Write a console application implementation of your algorithm.

CHALLENGE TASK 8.2

Combining validation checks

a Produce a flowchart and the pseudocode for a system that will ask for a 3-digit number below 501. The system will first check a value of three characters has been input. The system will then check the value is less than 501. The system will output the square of the input value. For example:

- An input of 1200 would fail the 3-digit criteria.
- An input of 550 would fail the below 501 criteria.
- An input of 250 would pass the criteria and the output would be 62500 (250^2).

b Write a console application implementation of your algorithm.

Type check validation

It is possible in Visual Basic to check specifically for numerical data types. The command checks the value provided and outputs a Boolean response to indicate if the input value is a number. While this is useful in determining if a value is numerical, it cannot distinguish between an integer and a real (decimal) number.

DEMO TASK 8.4

A system is required that asks the user to input the day of the month. The system should only accept integers as input.

Solution

Our code will need to identify the data type. It will then use a WHILE loop to check the input against the required criteria. If the data input is acceptable, the system will continue to run; if not, the system will output an error message to the user. Flowchart 8.4 shows the flowchart solution.

Here is the pseudocode solution:

```
DECLARE Number : INTEGER ← 0


OUTPUT "Enter a number"
INPUT Number


WHILE Number IS NOT Integer DO
    OUTPUT "Not integer,
    re-enter value"
    INPUT Number
ENDWHILE


Continue with program
```

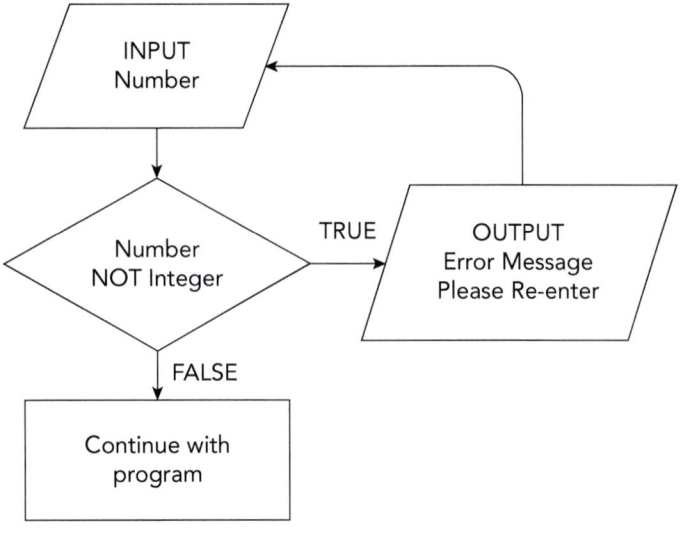

Flowchart 8.4: Flowchart pseudocode for type check

> **CONTINUED**
>
> The flowchart or pseudocode design of this type check validation is reasonably straightforward. However, the approach needed when coding is more complex. Although the user input will eventually be held in a variable with an integer data type, it is not possible for the user input to be placed directly into that variable. Before it can be stored in the integer variable the input must be checked to ensure it is a number. Failing to do this could result in a non-numerical value being passed to a variable designed to hold integer values. This would result in a system error.
>
> The solution is to follow several steps:
>
> 1. Take the input and store as a string variable.
>
> 2. Check if the value in the string variable is a number.
>
> 3. If the value is not a number reject the input.
>
> 4. Otherwise store the value as an integer variable.
>
> ```vb
> Sub Main()
>       Dim StringValue As String = ""
>       Dim NumberValue As Integer = 0
>       Console.WriteLine("Enter an integer value")
>       StringValue = Console.ReadLine()
>       'The IsNumeric inbuilt function. Takes the value in the variable
>       StringValue
>       'as a parameter input - The function will return TRUE if the value
>       equates to
>       'a number or FALSE if it does not equate to a number.
>       'Setting the loop criteria to FALSE will mean the loop will run and
>       request
>       ' re-entry while the value is not numerical. As soon as a number is
>       entered the
>       'loop will exit and the value in the string variable will be stored in
>       the
>       'integer variable
>       Do While IsNumeric(StringValue) = False
>          Console.WriteLine("Not an integer, re-enter value")
>          StringValue = Console.ReadLine()
>       Loop
>
>
>       NumberValue = StringValue
>
>
>       'Program code continues
>       Console.ReadKey()
>    End Sub
> ```
>
> The `IsNumeric()` function will return TRUE if any numerical value is tested. In this task, it was required that an integer value be entered. The `IsNumeric()` function will achieve effective validation if the requirement is only to check for a number that is then stored as a real value. However, it will not always achieve a reasonable solution to the requirement to only input an integer value.

> **CONTINUED**
>
> Any real (decimal) value that passes the `IsNumeric()` test and is then stored in an integer variable will be rounded. This may cause unexpected results, as shown in Table 8.2:
>
> | Input Value | Pass Test ? | Stored value in Integer variable | Result |
> |---|---|---|---|
> | Ten | Will be rejected | | |
> | 11 | Will be accepted | 11 | Stored value and input match. |
> | 11.3 | Will be accepted | 11 Rounded from 11.3 | Stored value and input value will be different but still represent the same integer element of 11. |
> | 11.6 | Will be accepted | 12 Rounded from 11.6 | Stored value and input value are different integer values. |
>
> **Table 8.2:** Results of real values in integer variables
>
> For many tasks, the rounding of a real input value would be sufficient, but there is a way to only accept integer values and reject all other inputs. A series of pre-programmed, pre-tested code is available to Visual Basic programmers to support these types of task. Figure 8.2 shows a WHILE loop condition that calls the `TryParse` function, a library routine that attempts to convert the input value into the data type indicated – in this case, a 32-bit integer value. If successful, an integer value is returned and stored in the required variable, and the Boolean value TRUE is returned to indicate success. If unsuccessful, the function does not store the value in the variable and the Boolean value FALSE is returned.
>
> ```
>                data type against which              variable to which the result
>                the check is to be made              of the parse is returned
> Do While Int32.TryParse(Console.ReadLine(), Number) = False
>                                 value which is parsed
>                                 to the data type
> ```
>
> **Figure 8.2:** A Try Parse command
>
> The following code make use of the Try Parse function:
>
> ```
>     Sub Main()
>         Dim Number As Integer = 0
>         Console.WriteLine("Enter an integer value")
>
>         Do While Int32.TryParse(Console.ReadLine(), Number) = False
>             Console.WriteLine("Not an integer, re-enter value")
>         Loop
>
>         'Program code continues
>         Console.ReadKey()
>     End Sub
> ```
>
> Using `TryParse(Console.ReadLine(), Number) = False` as the condition for the WHILE loop will cause the loop to continue while the conversion is unsuccessful. The loop will exit once a successful conversion has been made and the appropriate integer value has been returned to the variable `Number`.

> **PRACTICE TASKS 8.7–8.8**
>
> **8.7 Area of a rectangle**
>
> a   Produce a flowchart and the pseudocode for a system that will ask for the length and width of a rectangle to be input. The system will provide an error message if a non-numeric value is provided. Once the values have passed the criteria, the system will output the area of the rectangle.
>
> b   Write and test a console application implementation of your algorithm.
>
> **8.8 Number of people**
>
> a   Produce a flowchart and the pseudocode for a system that will take as an input the number of people attending a concert. The system will provide an error message if a non-integer value is provided.
>
> b   Write and test a console application implementation of your algorithm.

> **CHALLENGE TASK 8.3**
>
> **Month and year input**
>
> a   Produce a flowchart and the pseudocode for a system that will ask the user to input a month and year. Each value should be given as a numerical value. For example, January 2017 would be input as two integer values 1 and 2017.
>
> The system will need to check:
>
> i   Both inputs are input as integer values.
>
> ii The month input is in the range 1 to 12 inclusive.
>
> iii The year value must be greater than 2000.
>
> The system will provide appropriate error messages if these criteria are not met.
>
> b   Write and test a console application implementation of your algorithm.

Format check validation

> **DEMO TASK 8.5**
>
> *A form is required that asks the user to input the date in the format dd/mm/yyyy. The system will be required to check each element of the user input to check it matches the required pattern.*
>
> **Solution**
>
> The code will be required to check each element of the user input to ensure it matches a predetermined pattern. If the date matches the pattern, the system will continue to run; if not, the system will output an error message to the user. Flowchart 8.5 shows the flowchart solution:

CONTINUED

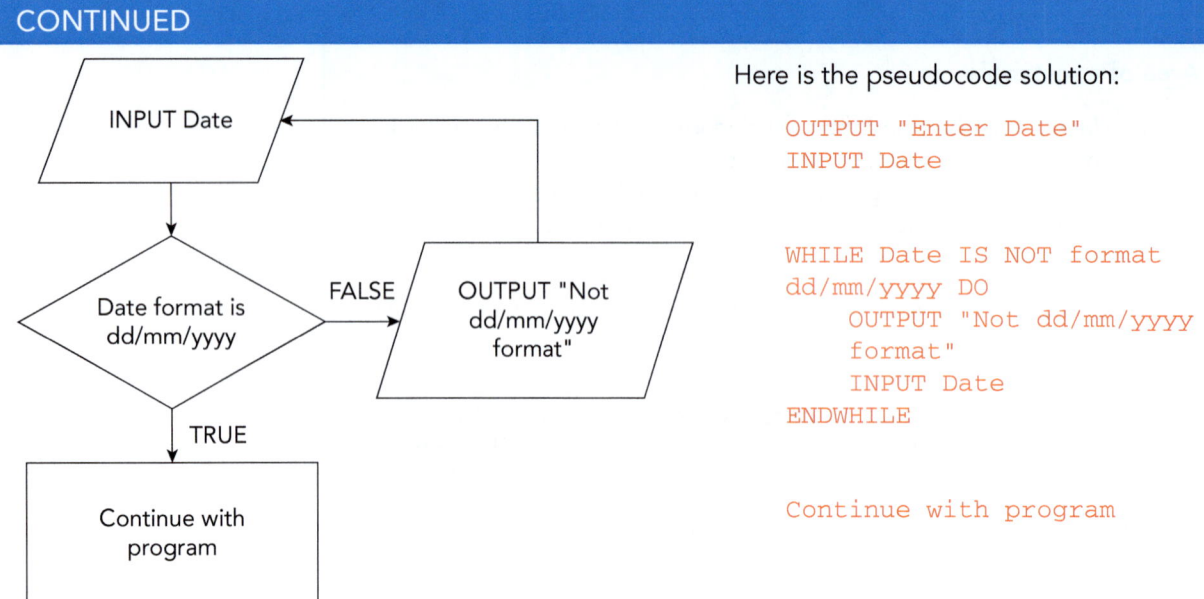

Here is the pseudocode solution:

```
OUTPUT "Enter Date"
INPUT Date

WHILE Date IS NOT format
dd/mm/yyyy DO
    OUTPUT "Not dd/mm/yyyy
    format"
    INPUT Date
ENDWHILE

Continue with program
```

Flowchart 8.5: Flowchart for format check

As with the type check, it is possible to ensure the correct format by using the `TryParse` method. The following example shows how `TryParse` has been used with the Date data type. Compare it with the Int32 data type that is used in the code in the Type check validation (earlier in this section) to check for integer values. Not only will `Date.TryParse` reject incorrect entries, it will also convert genuine dates to the expected date format. For example, if 7/1/20 was input, `Date.TryParse` would recognise the input as a genuine date and output the date in the correct format – 07/01/2020.

While this `Date.TryParse` feature is very useful, it does have some drawbacks. Consider what might happen if, when the user input 7/1/20, they had intended to input a date from the 1920s – 07/01/1920. Inputting the full date will avoid this error.

```vb
Module Module1
    Sub Main()
        Dim StringDate As String = ""
        Dim DateInput As Date
        Console.WriteLine("Enter date in the format dd/mm/yyyy")
        Do While Date.TryParse(Console.ReadLine(), DateInput) = False
            Console.WriteLine("Wrong format")
            Console.WriteLine("Enter date in the format dd/mm/yyyy")
        Loop
        Console.WriteLine(DateInput)
        StringDate = DateInput.ToString("dd MM yyyy")
        Console.WriteLine(StringDate)


        'Program code continues
        Console.ReadKey()
    End Sub
End Module
```

8 Checking inputs

> **TIP**
>
> Although the variable `DateInput` is declared as a Date data type, it is actually saved as a DateTime data type. When you run the code, you will see that the date output includes a time element. To avoid having the time displayed, it is possible to use the `ToString` command. This will convert the date to a string value that can be output. The way the date is shown in the string can be defined by including the required format in the brackets.
>
> Reproduce this code in a console application and see what happens when you use these `ToString` commands:
>
> ```
> StringDate = DateInput.ToString("dd/MM/yyyy")
> StringDate = DateInput.ToString("dddd dd/MMM/yyyy")
> StringDate = DateInput.ToString("dddd")
> StringDate = DateInput.ToString("yyyy")
> ```

> **PRACTICE TASK 8.9**
>
> **Birthday system**
>
> a Produce a flowchart and the pseudocode for a system that will ask for a person's date of birth. If the value input is not a valid date, the system will provide an error message. If the date is valid, the system will output the day the person was born.
>
> b Write and test a console application implementation of your algorithm.
>
> c Test your system to see if it can recognise leap years – try inputting the date 29 Feb for different years.

Check digit validation

> **DEMO TASK 8.6**
>
> A system is required that implements ISBN-13 check digit validation. Write the pseudocode and flowchart for this system.
>
> **Solution**
>
> Flowchart 8.6 shows an ISBN-13 check digit validation.
>
> 1 In ISBN-13 validation, the 13th digit is removed as this is the check digit.
>
> 2 All the other numbers are assigned a multiplier of 1 or 3, alternating from 1. These numbers are used as multipliers for their corresponding digits.
>
> 3 All the products are then added together to get the sum.
>
> 4 The modulus of an integer division of the sum of the products by 10 is found.
>
> 5 The modulus is then subtracted from 10. This should match the check digit if the ISBN number is correct.

> **CONTINUED**
>
> ISBN-13 check digit of 978-1-108-93567-8 is calculated as follows:
>
> 978-1-108-93567-?
>
> = 9×1 + 7×3 + 8×1 + 1×3 + 1×1 + 0×3 + 8×1 + 9×3 + 3×1 + 5×3 + 6×1 + 7×3
>
> = 9 + 21 + 8 + 3 + 1 + 0 + 8 + 27 + 3 + 15 + 6 + 21
>
> = 122
>
> 122 ÷ 10 = 12 remainder 2
>
> 10 − 2 = 8
>
> ? = 8

Flowchart 8.6: Flowchart for check digit validation of ISBN-13 numbers

CONTINUED

The ISBN number, taken as input, is a pseudo-number. This is input as a string and each digit is accessed by the string index (starting from zero). Manipulating string values is discussed in Chapter 11.

```vb
Module Module1
    Sub Main()
        Dim ISBN As String = ""
        Dim Multiplier As Integer = 1
        Dim Total As Integer = 0
        Dim CheckDigit As Integer
        Dim Modulus As Integer = 0


        Console.WriteLine("Enter ISBN Number")
        ISBN = Console.ReadLine()
        'Substring command used to store last digit in variable CheckDigit
        CheckDigit = ISBN.Substring(12, 1)


        'Loop through the first 12 digits in the ISBN number
        For Counter = 0 To 11
            'Keep a running total of the digit * the multiplier (either 1 or 3)
            'use the value of the loop counter in the substring command to
            iterate
            'through all the individual digits
            Total = Total + ISBN.Substring(counter, 1) * Multiplier
            'An IF statement to swap the value of multiplier between 1 and 3
            If Multiplier = 1 Then
                Multiplier = 3
            Else
                Multiplier = 1
            End If
        Next


        'Obtain the modulus (remainder) of an integer divison by 10
        Modulus = Total Mod 10


        'IF statement to check the value of Checkdigit (the 13th digit)
        'with the result of 10 - the modulus calculated
        If 10 - Modulus = CheckDigit Then
            Console.WriteLine("VALID")
        Else
            Console.WriteLine("INVALID")
        End If
        Console.ReadKey()
    End Sub
End Module
```

> **CAMBRIDGE IGCSE™ & O LEVEL COMPUTER SCIENCE: PROGRAMMING BOOK**

PRACTICE TASK 8.10

Barcodes

An EAN-8 barcode (used on small packages) consists of an eight-digit number where the first seven digits provide a code and the last digit is a check digit. The check digit is calculated, from the previous seven digits, in the following way:

Digit placement	1st	2nd	3rd	4th	5th	6th	7th	8th
Digit	6	1	1	7	2	3	8	2

- Add the digits in the odd-numbered positions (first, third, fifth, seventh) together and then multiply the answer by three. (For example, 6 + 1 + 2 + 8 = 17; 17 × 3 = 51.)

- Add the digits in the even-numbered positions (those in positions two four, and six) being careful *not* to include the check digit. (For example, 1 + 7 + 3 = 11.)

- Add the results of the two above calculations together and then find the modulus of the result of the addition result when integer divided by 10. This is the check digit. (For example, 51 + 11 = 62; 62 ÷ 10 = 6 remainder 2. The check digit is 2.)

a   Produce a flowchart and the pseudocode for a system that inputs the 8 digits from a barcode and outputs 'valid' or 'invalid' using the system just described.

b   Write and test a console application or Windows Forms application implementation of your algorithm.

Checking for specific characters

Although not a specific validation method, it is often required that input contains required elements. For example, a password might be required to include a capital letter and a symbol.

Visual Basic provides methods of checking for specific characters.

Contains command

The `contains` command will check that a string value contains a specific character. It returns the Boolean value TRUE if the sting value contains the specific character, or the Boolean value FALSE if the character is not present.

```
Sub Main()
    Dim StringValue As String = ""
    Console.WriteLine("Enter a value")
    StringValue = Console.ReadLine()
    If StringValue.Contains("£") = True Then
        Console.WriteLine("VALID")
    Else
        Console.WriteLine("INVALID")
    End If
    Console.ReadKey()
End Sub
```

This code shows how the `contains` command can be used to check if a string value includes a £ symbol.

The required character is provided in the brackets.

The command will only return TRUE or FALSE it can count the number of occurrences of the character.

Checking individual characters in the string

While the contains command is useful, it will only check for a specific character. It cannot, for example, check if the string includes a capital letter or any symbol. To achieve this, the code will need to check each character in the string individually.

A series of `Char.Is` commands are available in Visual Basic to check if a single character is of a defined type. This example of the `Char.Is` format shows the `Char.IsDigit` command:

```
Dim IndChar As Char
If Char.IsDigit(IndChar) = True Then
```

The command is passed the character value in the variable `IndChar` and will return either the Boolean value TRUE if the character is a digit (a number 0 to 9) or the Boolean value FALSE if it is not a digit.

Table 8.3 shows some of the `Char.Is` commands:

Command	Use
`Char.IsDigit(IndChar)`	Checks if the character passed is a digit (0 to 9).
`Char.IsLetter(IndChar)`	Checks if the character passed is a letter either (a to z) or (A to Z).
`Char.IsUpper(IndChar)`	Checks if the character passed is an upper-case letter (A to Z).
`Char.IsLower(IndChar)`	Checks if the character passed is an upper-case letter (a to z).
`Char.IsSymbol(IndChar)`	Checks if the character passed is a symbol (any symbol on a standard keyboard + many other special symbols).

Table 8.3: Some `Char.Is` commands

Using substring commands (see Chapter 11) to identify individual characters in a string combined with the `Char.Is` commands allows a programmer to check if a string contains either one of more characters of a specific type.

DEMO TASK 8.7

Consider the situation where a system is required to check if a password contains both uppercase and lowercase characters.

Solution

This could be approached using the following steps:

Step 1 Set up a loop to iterate through each character in the password.

Step 2a At each iteration, check if the character is an uppercase or lowercase letter.

Step 2b Change different Boolean variables to TRUE to record once an uppercase or lowercase letter is identified. Once the variable has been changed to TRUE, it won't need to change to FALSE for subsequent checks.

> **CONTINUED**
>
> **Step 3** At the end of the loop, check both Boolean variables. If both are TRUE, the password contains both uppercase and lowercase letters. If either is FALSE, the password does not contain the required mix of characters.
>
> The console application code to achieve this task could be:
>
> ```vb
> Module Module1
>     Sub Main()
>         Dim Password As String = ""
>         Dim UCase As Boolean = False
>         Dim LCase As Boolean = False
>         Console.WriteLine("Please enter password")
>         Password = Console.ReadLine()
>
>
>         'Loop to iterate though all characters in password
>         'Loop limited to Password.Length -1 as string index is zero based
>         'so a password of 11 characters would be indexed 0 to 10
>
>         For Counter = 0 To Password.Length - 1
>            Dim IndChar As Char
>             'Obtain the individual character using substring command
>             'The loop counter value in the substring will iterate through all
>             characters
>             IndChar = Password.Substring(counter, 1)
>             'Use the Char.Is commands to identify Uppercase or Lowercase
>             'and change appropriate Boolean variable to TRUE
>             If Char.IsUpper(IndChar) = True Then
>                 UCase = True
>             ElseIf Char.IsLower(IndChar) = True Then
>                 LCase = True
>             End If
>         Next
>
>
>         'After loop has exited check value of Boolean variables
>         'and output appropriate message
>         If UCase = True And LCase = True Then
>             Console.WriteLine("VALID PASSWORD")
>         Else
>             Console.WriteLine("INVALID PASSWORD")
>         End If
>         Console.ReadKey()
>     End Sub
> End Module
> ```

SUMMARY

Accuracy of data entry is an important consideration in system design. Inaccurate data can lead to inaccurate outputs.

Validation is a technique in which the system checks data input against a set of predetermined rules.

Validation can identify obvious errors by detecting data that fails to meet the validation rules.

Validation is able to ensure that data input is reasonable but cannot guarantee data accuracy.

Six main forms of validation are used to check data as it is input:
- Presence checks ensure that data has been input.
- Range checks ensure that data falls within a predetermined range of values.
- Length checks ensure that data inputs contain a predetermined number of characters.
- Type checks ensure that data input is of a certain data type.
- Format checks ensure that data input meets a predetermined format, such as dd/mm/yyyy.
- Check digits are calculated from numerical data such as a barcode and added to the end of the data.

It is possible to check that specific characters are present in a string. This can be used to either accept only if the required characters are used or to reject if unacceptable characters are included.

Verification checks the integrity of data when it is entered into the system. This is often completed by the individual inputting the data.

Two common methods of verification are:
- Checking the input data against the original document or record.
- Double entry in which the data is entered twice and the entries compared to identify differences.

END-OF-CHAPTER TASKS

1. **a** Produce in pseudocode or as a flowchart, an appropriate algorithm for a system that takes a string value. The system will check that the string value is more than 10 characters and contains no numerical digits.
   b Write and test a console application or Windows Forms application implementation of your algorithm.

2. **a** Produce in pseudocode or as a flowchart, an appropriate algorithm for a system that takes as an input an integer value. The system will check that the input is an integer value greater than 16 and less than 100.
   b Write and test a console application or Windows Forms application implementation of your algorithm

3. A business sells and delivers bricks to builders. The minimum order the business will accept is 200 bricks. The maximum load they can deliver 6000 bricks. When the order is placed, the business also take the 5-digit CustomerID number of the builder. CustomerID numbers are numerical and begin with a 9. For example, 92311.
   A system is required to validate the data input is required.
   a Decide on appropriate validation rules for the system.
   b Produce in pseudocode or as a flowchart, an appropriate algorithm for your validation rules.
   c Write and test a console application or Windows Forms application implementation of your algorithm.

CONTINUED

4 A system is required to validate passwords. The user is required to enter the password twice and both entries must match.

The password must meet the following criteria:
- It must have a minimum of 6 characters and a maximum of 12 characters.
- It must contain at least three of these character types:
  - an uppercase letter
  - a lowercase letter
  - a symbol
  - a number.
- The system will validate the password entered and will output one of four messages:
  - Password accepted.
  - Passwords input do not match.
  - Password does not meet length requirements.
  - Password does not contain enough character types.

a Produce in pseudocode or as a flowchart, an appropriate algorithm for this system.

b Write and test a console application or Windows Forms application implementation of your algorithm.

Chapter 9
Testing

IN THIS CHAPTER YOU WILL:

- learn about the importance of testing systems
- learn how to identify logical, syntax and runtime errors
- learn how to dry run algorithms using trace tables
- learn how to identify appropriate valid, invalid and boundary data when testing systems.

Introduction

In common with many products, it is important to make sure systems work as expected before they are released to the final user. The complexity and critical nature of the system will determine how much testing is to be completed. For example, the computerised air traffic control system at an airport is more critical than a smartphone game and, therefore, will have undergone far more extensive testing as failure could be catastrophic.

There are several notable examples of disasters caused by poor testing. The destruction of the unmanned *Ariane 5* space rocket, due to the failure of untested code, is one of the most costly examples ever. The financial losses were measured in billions of dollars. An article published in *The New York Times* magazine in December 1996 has information about the software error that caused the disaster. You will be able to find the article if you search online for 'Ariane disaster'.

9.1 When to test

Testing can be broken into two distinct areas: alpha testing and beta testing.

Alpha testing

Alpha testing is completed during the programming of a system. It checks that each part of a program works as expected before being combined to make a complete system. Testing during the programming stage can also help trace the source of unexpected outcomes.

Beta testing

Beta testing is the formal testing that takes place once the system has been completed to ensure that the whole system meets expectations. The product that has been developed is installed on the systems where it is intended to be used. The normal users will complete a series of tests to ensure the system works in the target environment before it is finally released for general usage.

9.2 Debugging

Debugging is the process of detecting faults that cause errors in a program. This can be achieved by observing error messages produced by the IDE or by investigating unexpected results. The types of error that can occur are divided into three groups: logical errors, syntax errors and runtime errors.

Logical errors

Logical errors are errors in the design of the program that allow it to run but produce unexpected results. They can result from the use of an incorrect formula or the incorrect use of control structures such as IF statements or loops. Examples include IF statements with incorrect conditions, or loops that iterate the wrong number of times.

> **KEY WORDS**
>
> **alpha testing:** early testing that takes place during the programming of a system.
>
> **beta testing:** formal testing once the system has been completed.
>
> **debugging:** a general term for systematically searching for problems in programs that are not working properly and fixing them.
>
> **logical errors:** result in code that runs but produces unexpected results.

Logical errors are also caused by implementing an incorrect sequence of statements, such as performing a calculation before assigning values to the variables.

Logical errors usually do not produce error messages. The problem is with the logic of the code not the execution of the code.

> **TIP**
>
> One source of logical errors is copying code from one part of your system to another part of the system. While this might reduce the time it takes to retype similar code, it can be the source of errors if variables or boundaries are included in the copied section. Forgetting to change the original variables or boundaries will mean the code will still run but you are likely to get unexpected results.

Syntax errors

Syntax errors are errors in the use of the programming language, such as incorrect punctuation or misspelt variables and control words. Examples include IF statements with missing colons or incorrect use of assignment. The IDE will usually generate error messages indicating the reason for the error.

Runtime errors

Runtime errors are errors that are only identified during the execution of the program. They can result from mismatched data types, overflow or divide-by-zero operations.

Data type errors include:

- Passing String data to an Integer variable, which will probably cause the system to crash.
- Passing Real data to an Integer variable; the variable will round the input data to the nearest whole number – the system will execute the code but produce unexpected results.

Overflow errors occur when the data passed to a variable is too large to be held by the data type selected. In the theory element of the syllabuses, you will have used this term to describe a situation where a nine-bit binary number is stored in an eight-bit byte. This can often result from calculations during the execution of a program. For example, in some programming languages, the data type `Short` can be used to hold numbers between −32 767 and +32 767. If a variable of this data type were assigned the result of the square of any number greater than 182, it would produce an overflow error.

In mathematics, it is not possible to divide by zero because any number can be divided by zero an infinite number of times. If a program includes a division calculation that divides by a variable holding the value zero, the system will produce an error. Figure 9.1 shows a sample of code that attempts to divide by zero and the associated error message.

> **KEY WORDS**
>
> **syntax errors:** mistakes made in the code equivalent to spelling and punctuation mistakes in English.
>
> **runtime errors:** problems with the code that only become evident when the program is run, for example, attempting to divide by zero.
>
> **overflow errors:** occur when the data passed to a variable is too large to be held by the data type selected.

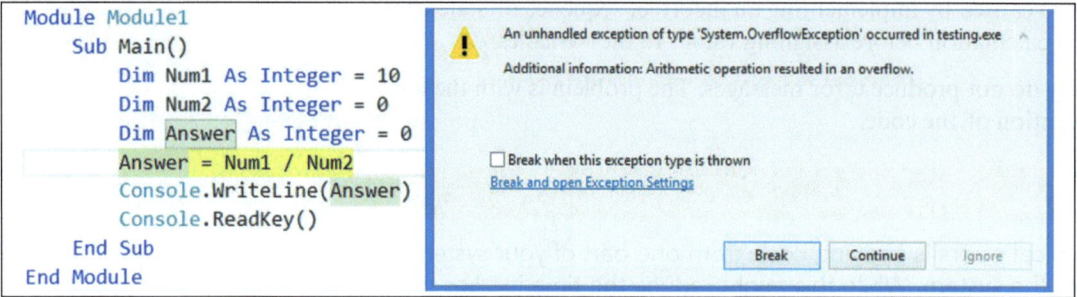

Figure 9.1: Overflow error message resulting from a divide by zero runtime error

9.3 IDE debugging tools and diagnostics

Many IDEs include sophisticated **diagnostics** that help identify possible bugs and provide the user with useful error messages. These tools are only able to identify errors in the code, not in the logic of the code, and as a result are unable to identify logical errors Visual Studio Express provides debugging support for both syntax and runtime errors.

> **KEY WORD**
>
> **diagnostics:** the systematic process of trying to diagnose what is wrong with a program.

Syntax diagnostics

Syntax errors are identified and underlined during the process of typing the code. Supporting error messages are available when the curser is placed over the underlined code. An error window also provides information about errors, including the line number of the error. Figures 9.2–9.4 show three examples of the type of error messages given by the IDE.

```
Module Module1
    Sub Main()
        Dim Input1 As Integer = 10
        Dim Input2 As Integer = 10
        Input1 = Console.Reedline
        Input3 = Console.ReadLine
        Input2 = "Hello"
        If Input1 = 10 Then
                'If' must end with a matching 'End If'.
                Show potential fixes (Ctrl+.)
```

Figure 9.2: Error message provided when curser is placed over underlined code

```
Input1 = Console.Reedline
Input3 = Console.ReadLine
Input2 = "Hello"       Change 'Reedline' to 'ReadLine'.   ⊗ BC30456 'Reedline' is not a member of 'Console'.
If Input1 = 10 Then
                                            Dim Input2 As Integer = 10
                                            Input1 = Console.Reedline
                                            Input1 = Console.ReadLine
                                            Input3 = Console.ReadLine
                                            ...
                                            Preview changes
```

Figure 9.3: Correction suggestions offered when drop down selected

188

9 Testing

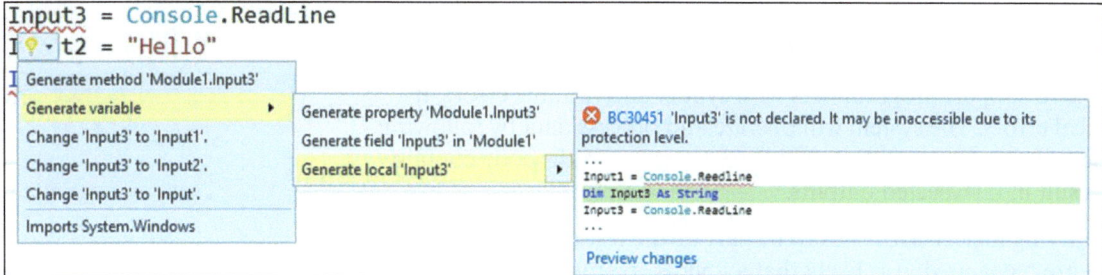

Figure 9.4: Correction suggestions where error may have multiple solutions

In addition to the in-code syntax checking, Visual Studio Express also includes an error list. This will show all the errors identified in the code. Double clicking each error will take you directly to the incorrect line in the code. Figure 9.5 shows an error list window.

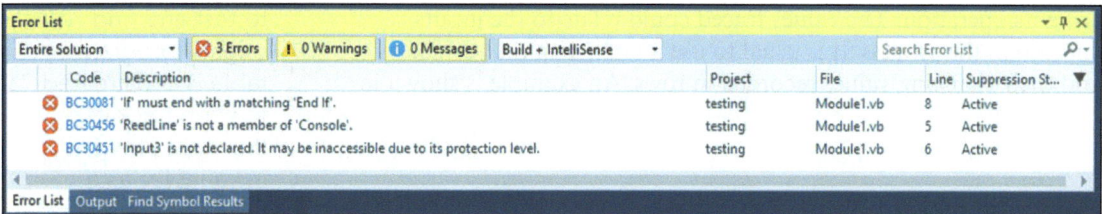

Figure 9.5: The error list window

Runtime diagnostics

Runtime errors are detected during execution of a program and specific error messages are provided. The code shown in Figure 9.2 declares the variable `Input2` with an integer data type `Dim Input2 As Integer`. But then attempts to place a string value in the variable with the code `Input2 = "Hello"`. This is an example of a runtime error. The IDE is unable to identify the potential error until the code executes and the data type mismatch is recognised. At that time, the code will halt and a runtime error message will be generated (as shown in Figure 9.6).

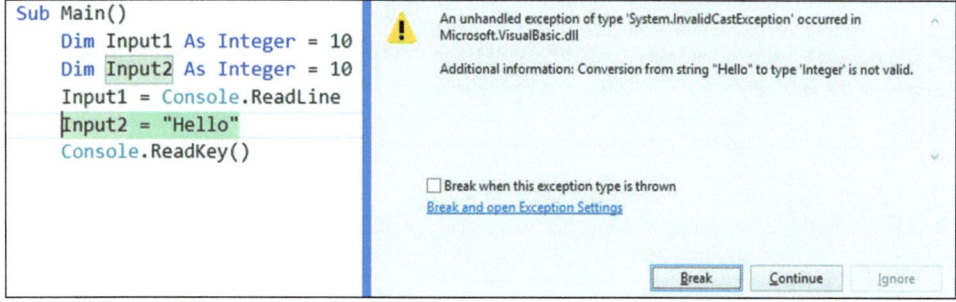

Figure 9.6: Runtime error message

189

9.4 Identifying logical errors

While the IDE is able to support programmers with syntax and runtime errors, it cannot identify logical errors. The system will operate and process data by following the code that has been written – the IDE is unable to determine if the code contains logical errors that result in unexpected outputs.

The process of identifying logical errors has to be part of the testing process. When unexpected outputs are recognised, it is likely that a logical error will be present in the code. The actual error will also have to be identified manually.

9.5 Dry running

Dry running is the process of working through a section of code manually to locate logical or runtime errors. This type of testing often uses a **trace table** to record values within a system during its operation. The values traced could relate to the inputs, outputs, or variables used in the process. It is usual to use a table with the variables listed as columns and their changing values recorded in rows. An example is shown in Demo Task 9.1.

> **KEY WORD**
>
> **trace tables:** a way to test and find bugs in programs. A table is constructed to keep track of the values held in variables as a program is stepped through line by line.

Tracing pseudocode

> **DEMO TASK 9.1**
>
> **Trace table**
>
> *A student has been challenged to write code that will calculate the integer division of X by Y without using the quotient operator. They have decided that an algorithm that counts how many times Y can be subtracted from X will produce the required outcome.*
>
> **Solution**
>
> Code snippet 9.1 is the pseudocode for their algorithm to calculate the integer division of X by Y.
>
> ```
> DECLARE X : INTEGER ← 0
> DECLARE Y : INTEGER ← 0
> DECLARE Count : INTEGER ← 0
> OUTPUT "Enter a value for X"
> INPUT X
> OUTPUT "Enter a value for Y"
> INPUT Y
> WHILE X > Y DO
>     X ← X - Y
>     Count ← Count + 1
> ENDWHILE
> OUTPUT Count
> ```
>
> Code snippet 9.1

> **CONTINUED**
>
> The trace table shown (Table 9.1) traces the execution of this algorithm. In this case, the algorithm has input values of X = 20 and Y = 6. Comments have been added to help explain the trace table. Comments are not normally required in a formal trace table.
>
> | X | Y | count | Output | Comments |
> |----|---|-------|--------|----------|
> | 0 | 0 | 0 | | Initialisation value. |
> | 20 | 6 | 0 | | The new values are input. |
> | 14 | 6 | 1 | | X is reduced by 6, Count is incremented by 1. LOOP returns to the WHILE condition check. As X > Y, the loop continues to run. |
> | 8 | 6 | 2 | | X is reduced by 6, Count is incremented by 1. LOOP returns to the WHILE condition check. As X > Y, the loop continues to run. |
> | 2 | 6 | 3 | | X is reduced by 6, Count is incremented by 1. LOOP returns to the WHILE condition check. As X < Y, the loop exits. |
> | | | | 3 | The value in Count is output. |
>
> **Table 9.1:** Trace table for the X/Y algorithm

> **PRACTICE TASK 9.1**
>
> **Trace table**
>
> The pseudocode algorithm (Code snippet 9.1 in Demo Task 9.1) contains a logical error. Complete a trace table with the input values of X = 24 and Y = 6 to identify the error.

Tracing a flowchart

DEMO TASK 9.2

Flowchart trace table

Flowchart 9.1 shows the flowchart for an algorithm. Table 9.2 shows the trace table which has been completed where the input value A = 2.

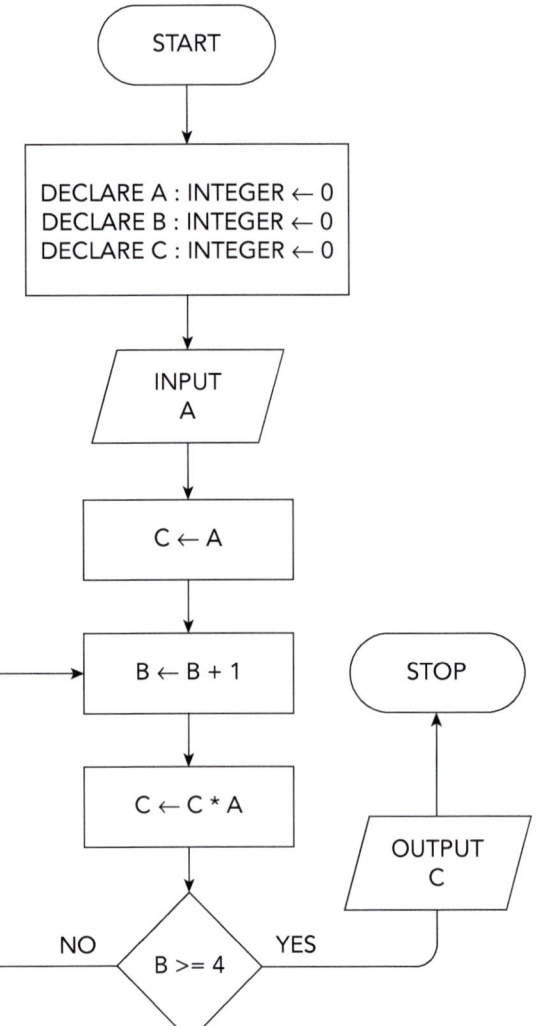

Flowchart 9.1: Flowchart for an algorithm

A	B	C	Output	Comments
0	0	0		Initialised values
2	0	0		Input A
2	1	2		C ← A
				B ← B + 1
2	1	4		C ← C * A
				Will loop as B < 4
2	2	8		B ← B + 1
				C ← C * A
				Will loop as B < 4
2	3	16		B ← B + 1
				C ← C * A
				Will loop as B < 4
2	4	32		B ← B + 1
				C ← C * A
				Loop exited as B = 4
2	4	32	32	Output value in C

Table 9.2: Trace table for an algorithm

PRACTICE TASK 9.2

a   Produce a trace table for the algorithm in Demo Task 9.2 where A = 3.

b   What is the aim of the flowchart in Flowchart 9.1?

c   What kind of loop is being suggested here?

TIP

To help with Practice Task 9.2b, you could complete a trace table for other values of A and use the outcome to identify what this algorithm is achieving.

9.6 Breakpoints, variable tracing and stepping through code

Although an IDE cannot identify logical errors, they do provide tools that assist programmers with this manual process. Visual Studio Express, in common with many IDEs, provide the programmer with the ability to run the program one line at a time, displaying the values held in the variables at each step. The programmer is helped when checking particular sections of their code by being able to run the program, as normal, until it meets a 'breakpoint'. These are inserted into the code by the programmer and will cause the system to run one line of code at a time once the breakpoint is reached.

To insert a breakpoint, right click on the line of code where you want to insert the breakpoint (see Figure 9.7).

> **KEY WORD**
>
> **breakpoint:** points that are inserted into the code by the programmer. They will cause the system to run one line of code at a time once the breakpoint is reached. Helps the programmer to track the value of variables as the program executes.

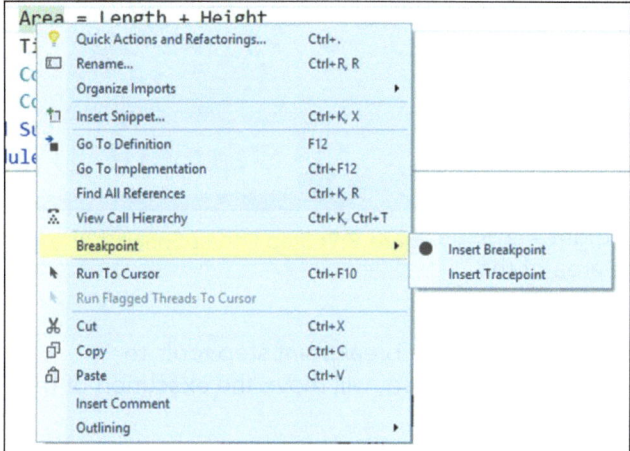

Figure 9.7: Inserting a breakpoint

Once inserted the line of code will show a red circle (see Figure 9.8).

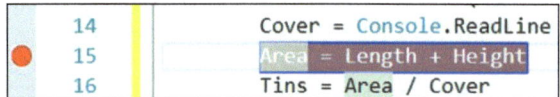

Figure 9.8: Breakpoint in code

> **DEMO TASK 9.3**
>
> This code has been written to calculate the number of tins of paint required to cover a wall. The code needs to be tested for errors.
>
> ```
> Dim Length As Decimal = 0
> Dim Height As Decimal = 0
> Dim Area As Decimal = 0
> Const Cover As Integer = 15
> Dim Tins As Integer = 0
> ```

CONTINUED

```
Console.WriteLine("Enter length in metres")
Length = Console.ReadLine()
Console.WriteLine("Enter height in metres")
Height = Console.ReadLine()

Area = Length + Height
Tins = Area / Cover
Console.WriteLine(Tins & " Tin(s) needed")
Console.ReadKey()
```

Solutions

The constant `Cover` indicates the maximum number of square metres that can be covered by one tin of paint. The user inputs the length and height of the wall and the system should output the number of tins of paint the painter needs to buy. The algorithm does not produce the expected results, so the programmer decides to use breakpoints to help identify the error. The programmer is happy that the inputs are correct, so they insert the breakpoint after the input sequence.

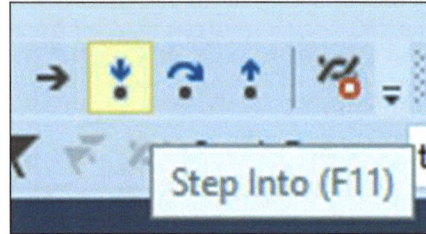

Figure 9.9: Step Into option

To test the system, they use the values length = 8.0 and height = 2.5. They expect the system to output '2 tin(s) needed', as the area of the wall will require 1.33 tins of paint.

When the code is run, it stops at the breakpoint. The programmer uses the breakpoint step tools to navigate through the code. The 'Step Into' option (Figure 9.9), when pressed, will move the execution of the code to the next line.

When using breakpoints, the current value of variables is shown by hovering over the variable name. They are also shown in the 'Locals' window, which is normally below the main code windows (Figure 9.10).

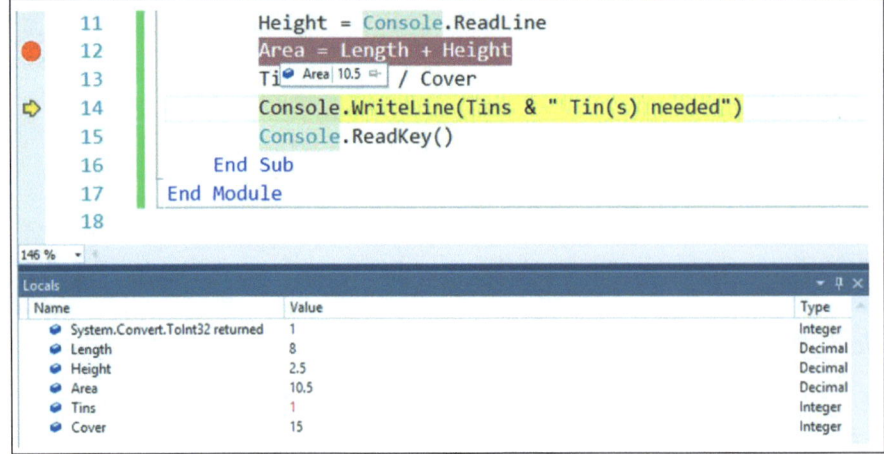

Figure 9.10: Using breakpoint to view the value contained by the local variables

It can now be seen that the value of the variable `Area` is 10.5, not the expected value of 20. The error is likely to be in the calculation, and on closer investigation, it can be seen that the values of `Length` and `Height` have been added rather than multiplied.

PRACTICE TASK 9.3

Once the calculation error of the variable `Area` was corrected (Demo Task 9.3), the programmer continued to receive unexpected results for some input values.

a Reproduce the code and use a breakpoint and a range of values to identify the second error.

b How could this error be corrected?

CHALLENGE TASK 9.1

Being able to spot errors in other programmers' code is a useful skill. You can practise this skill by working with other students.

Each student should deliberately introduce errors into a section of code. The code will need to be supported by an explanation of what the system was intended to achieve. Logical errors are often the hardest type of error to spot. Challenge each other to spot and correct the errors in the code.

9.7 Beta testing

A formal test schedule is designed to test all possible events that a system could experience. It will test normal expected operation as well as extreme inputs or usage. The test schedule will identify the elements of the system to be tested and the data to be used in the tests. Each set of test data and the expected outcome is known as a 'test case'. The data used will fall into three categories, as described in Table 9.3. The example data in Table 9.3 is based on a system designed to calculate the grade achieved by students in an examination. The inputs are the students' marks and the maximum possible mark.

Type of test	Description	Example data
Normal data	Data that is expected to be met in the normal operation of the system. It meets the expected validation rules. The system should produce the expected outcome.	Integer values between zero and the maximum possible score.
Abnormal data	Data that will not form part of the expected input range. The system should reject the data and output appropriate error messages.	Non-Integer values (it is not possible to get half a mark). Values less than zero or more than the maximum score. Textual inputs, such as 'TEN'.

(continued)

Type of test	Description	Example data
Extreme data	Data that is at the very limit of the acceptable values. This could be either a maximum or minimum value depending on boundaries set.	Data that falls at grade boundaries. For example: • The maximum mark is 100. • The grade boundary for a B is >=70% but < 80%. Data at maximum boundary: 79 Data at minimum boundary: 70
Boundary data	Data that is that is on the edge of being accepted and being rejected.	Data that falls either side of grade boundaries. For example: • The maximum mark is 100. • The grade boundary for an A is 80%. Data at the boundary: 80 Data one mark below the grade boundary: 79

Table 9.3: Data categories for beta testing

> **TIP**
>
> It is common to see data values that fall into several of these data categories. In the example shown in Table 9.3, the value 79 represents valid data, extreme data when testing the B grade or boundary data when testing either the B or A grade.

DEMO TASK 9.4

An algorithm has been designed to calculate weekly payment due to employees at a factory. It takes as input the number of hours an employee has worked and uses the rules shown to calculate and output the payment due. Create a test schedule to check the algorithm works as expected.

Standard hourly rate payable for the first 35 hours worked in the week.	$14.50
Overtime rate payable for each hour over the first 35 worked in the week.	$21.50
The maximum number of hours an employee is permitted to work in a single week is 50. Employees only work in full hour or half hour segments.	

CONTINUED

Solution

The following is an example of a test schedule that could be used to test the algorithm:

Test	Area to Test	Data to be used	Expected Outcome
1	Input validation Abnormal data	53 hours worked	Input will be rejected as over the maximum 50 hours.
2	Input validation Abnormal data	Negative value for hours such as −20	Input will be rejected not able to work negative hours.
3	Input validation Abnormal data	40.75 hours worked	Input will be rejected as only half hour segments allowed.
4	Boundary test	35 hours worked	Output $507.50
5	Boundary test	36 hours worked	Output $529.00
6	Boundary test	50 hours worked	Output $830.00
7	Normal Data	39 hours worked	Output $593.50

PRACTICE TASK 9.4

Beta testing

A children's address book app holds the name, telephone number and date of birth of each friend entered. The user can input a date and the system will output, the age, name and telephone number of each friend who has a birthday on the date entered.

For each of the data items, decide on:

a   the appropriate data type

b   appropriate input validation that could be applied

c   abnormal and, where appropriate, boundary data that could be used to test input validation.

CHALLENGE TASK 9.2

A system is required by a teacher to calculate the average height of a group of students. The teacher will input the height of the students in the group. The input sequence will be ended with an input of −1. The system will output the average height of the group.

a   Create a test schedule that could be used to test the algorithm required for this task.

b   Write a console application implementation of the algorithm and use your test schedule to check the algorithm for errors.

SUMMARY

It is important to test systems to ensure they will perform as expected.
Alpha testing is completed during the programming of a system.
Beta testing is formal testing once the system has been completed.
Logical errors are errors in the logic of the process performed by the code. The code will run but will produce unexpected results. These need to be debugged manually using, for example, trace tables.
Syntax errors are errors in the syntax used within the code. It is likely that these will be identified by the IDE diagnostics.
Runtime errors only appear during the execution of the code. Attempting to divide by zero is a common runtime error.
IDEs identify syntax errors and runtime errors. IDEs also provide useful tools to help debug logical errors by providing the ability to step through code and add breakpoints.
Trace tables provide a useful means of tracing the value of variables, inputs, and outputs at each step of an algorithm. They can be helpful in identifying logical errors.
Normal data is met by the system in its normal operation.
Abnormal data is data that the system is not expecting. The system should identify and reject abnormal data and provide appropriate error messages.
Boundary data is data that falls either side of a boundary. The largest acceptable value and the smallest value that should be rejected. It is used to check the logic of the program at boundaries.
Extreme data is data that is at the extremes of acceptable data. The largest and smallest acceptable values.

END-OF-CHAPTER TASKS

1 In the following two scenarios, say which is *alpha testing* and which is *beta testing*?
   a A programmer has completed programming the first level of a new game. She then runs some tests to check that her code is working before moving on to the other levels.
   b After completing a weather app, a software company sends early copies of the software to a group of people to try it out for a month and report back any problems they find.

2 In the following three scenarios, what are the most likely kinds of errors to look for: *logical errors*, *syntax errors*, or *runtime errors*?
   a A program is undergoing beta testing and it is discovered that, after a little while, the program freezes because it has run out of allocated memory.
   b A program refuses to run because of a missing speech mark.
   c A program is being written to calculate the total cost of products brought to a till at a supermarket. During alpha testing, a strange thing is happening: No matter what products are in the basket, the total cost is £99.99.

3 In the three scenarios given in Task 2, what are the best ways of trying to diagnose and fix the problems encountered?

CONTINUED

4 The following algorithm is designed to accept a series of numbers in the range 1 to 100, with the sequence being ended by the user inputting a negative number. At the end of the sequence, the system will output the smallest number input, the largest number input, and the average of the numbers input. The algorithm contains a number of logical errors.

Identify the errors **and** explain how they could be corrected.

```
DECLARE Small : REAL ← 0   DECLARE Large : REAL ← 0   DECLARE Total : REAL ← 0
DECLARE Counter : INTEGER ← 0   DECLARE Number : REAL ← 0
DECLARE Average : REAL ← 0
OUTPUT "Enter Number"
INPUT Number
WHILE Number <>-1 DO
    IF Number >0 OR Number < 100
      THEN
        IF Number < Small
          THEN
            Small ← Number
        ENDIF
        IF Number > Large
          THEN
            Number ← Large
        ENDIF
        Total ← Total + Number
        OUTPUT "Enter Number"
        INPUT Number
        Counter ← Counter + 1
        Average ← Counter / Total
        OUTPUT Average
      ELSE
        OUTPUT "Out of range - enter number"
        INPUT Number
    ENDIF
ENDWHILE
OUTPUT "Small"
OUTPUT Large
```

> **TIP**
>
> There are nine different logical errors to identify.

5 Write and test a console application implementation of your corrected algorithm from Task 4.

Chapter 10
Arrays

IN THIS CHAPTER YOU WILL:

- know how to define an array using flowcharts and pseudocode
- declare and use a one-dimensional array
- declare and use a two-dimensional array
- be able to read and write values from an array
- learn how you can use arrays to organise data
- sort an array using the inbuilt sort command
- understand how to code a bubble sort algorithm.

10 Arrays

Introduction

A common way of recording many data items is to make a list. Many people will record all the tasks they hope to achieve in a 'To-Do' list. Scientists may record their observations and measurements of a chemical reaction as a timed list. Teachers are likely to have a list of students in the class. The reason that lists are so common is that they record many data items related to a single event in one convenient place.

Arrays provide programmers with the equivalent of a list. Arrays are a series of related data items all conveniently stored in one location.

10.1 What is an array?

An **array** is a data structure that can hold a set of data items under a single identifier. In the same way that a variable holds data of a specific data type and has a name that can be used to identify it, an array also holds data of a specific type and can be referred to by its name or label. The major difference is that while a variable can only hold one data value, an array can hold multiple values. For example, if you wished to store the surnames of 25 students using variables, you would have to declare 25 variables. Using an array, you could store all 25 values in a single array set up to accept 25 data items.

10.2 Declaring a one-dimensional array

A **one-dimensional array** is often visualised as a single row of data. Each of the data items is identified by a single index number. Later in this chapter (Section 10.7), we will look at two-dimensional arrays that can store multiple rows of data and are indexed by two index values. Many people visualise these as a grid with row and column index values.

Declaring a one-dimensional array is a similar process to declaring a variable – the same naming and data type requirements exist. The difference is that you need to define the size of the array, which will be determined by the number of data items that the array is required to hold. Each individual value held within an array is identified by an **array index** number (see Table 10.1). Index numbers are sequential. In Visual Basic, the numbering starts from zero.

Index number	0	1	2	3	4
Data item	Alpha	Beta	Gamma	Delta	Epsilon

Table 10.1: A diagrammatic representation of an array designed to hold the first five letters of the Greek alphabet

The pseudocode format for declaring an array capable of holding the five values is as follows. NOTE: Observe how the range of indices is 0 to 4:

```
DECLARE GreekLetter : ARRAY [0 : 4] OF STRING
```

The first index (0) is also known as the lower bound and the last index (4) is also known as the upper bound.

> **KEY WORDS**
>
> **array:** a variable that can hold a set of data items of the same data type under a single identifier.
>
> **one-dimensional array:** a linear array with a single index set. Contains one row of data with multiple elements; each element is identified by a unique index number.
>
> **array index:** a series of sequential numbers that reference the data items in an array. In Visual Basic, the first index value is zero. This is known as a zero-based index

When using pseudocode to declare an array is it important that you explicitly state the first index (lower bound) and last index (upper bound).

In this book, the first index will always be zero as that matches the indexing of an array in Visual Basic. However, in some systems, the first index can default to one. By explicitly stating both the first (lower bound) and last index (upper bound), your logic can be easily understood.

Examples where the array is numbered from 1 will also have the final index increased by 1:

```
DECLARE GreekLetter : ARRAY [0 : 4] OF STRING
```
  – Array indexed from 0 has five locations (0,1,2,3,4).
```
DECLARE GreekLetter : ARRAY [1 : 5] OF STRING
```
  – Array indexed from 1 has five locations (1,2,3,4,5).

Whichever method is used, it is crucial that you remain consistent when declaring and using an array. It is not uncommon to see students declare an array using one approach to indexing an array but then produce code based on a different indexing approach.

Later in this chapter you will learn that a loop can be used to iterate through all the index positions in an array. Consider the situation where an array was indexed from 0 but the loop started from 1. The first index location would never be accessed by the loop.

The syntax for declaring a one-dimensional array in Visual Basic is shown in Figure 10.1:

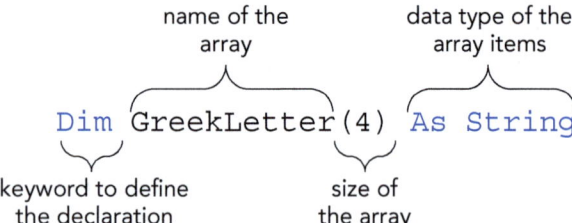

Figure 10.1: Visual Basic declaration syntax for a one-dimensional array

This declares an array that will hold five values.

In Visual Basic, the size of the array is defined by stating the final index number. The first index position of 0 is assumed. The range of this array is [0:4], representing five individually numbered data items.

10.3 Initialising arrays

While it is common to initialise a variable with a default value, it is unusual to initialise an array unless you want to place specific data items in the array at the time of declaration. If you wish to initialise an array in Visual Basic, the size of the array is defined by the number of data items used to initialise the array. The explicit reference to size – (5) – has to be omitted from the ( ) by removing the number:

```
Dim GreekLetter() As String = {"Alpha", "Beta", "Gamma", "Delta", "Epsilon"}
```

The range of this array is [0:4], defined by the five individually named data items.

10.4 Using arrays

Arrays offer programmers advantages over variables. Arrays allow many data items to be stored under a single identifier, and they give the programmer the ability to reference any individual data item by the appropriate array index. Iteration can then be used to complete read, write or search operations by looping through the data items. This makes arrays particularly effective when working with large data records.

Reading and writing data items

To read a data item, you reference it by the array name and the index number. For example, `GreekLetter(2)` holds the data item 'Gamma' (remember that the array index starts at 0, so index number 2 is the third item). The same logic applies when writing values to an array. The following code would write the letter 'C' to the item at the specified index position replacing the original data item:

```
GreekLetter [2] ← "C"
```

Now our array of Greek letters would read 'Alpha', 'Beta', C, 'Delta', 'Epsilon'.

> **SKILLS FOCUS 10.1**
>
> **OVERALL SYSTEM DECISIONS**
>
> **Integer array**
>
> Declare an array named 'Task' that is capable of holding four integer values. Write code to allow the user to input a data item to selected array positions. Write code to allow the user to output the value held in a selected array position.
>
> Although both tasks seem straightforward there are some possible complications. The designs in Flowchart 10.1 and the pseudocode show the INPUT and OUTPUT tasks as two individual tasks. This is accurate in showing the detail of the individual tasks but does not include consideration of how the user would indicate whether they wanted to INPUT or OUTPUT values.
>
> The second consideration is how the user would indicate the index they wished to use. As programmers, it is normal to consider the first array index to be 0. But users are unlikely to be familiar with this and concept and may well indicate the first index as index 1.
>
> **Step 1** Design the underlying processes (see Flowchart 10.1 (a and b) and the pseudocode).

> **CONTINUED**

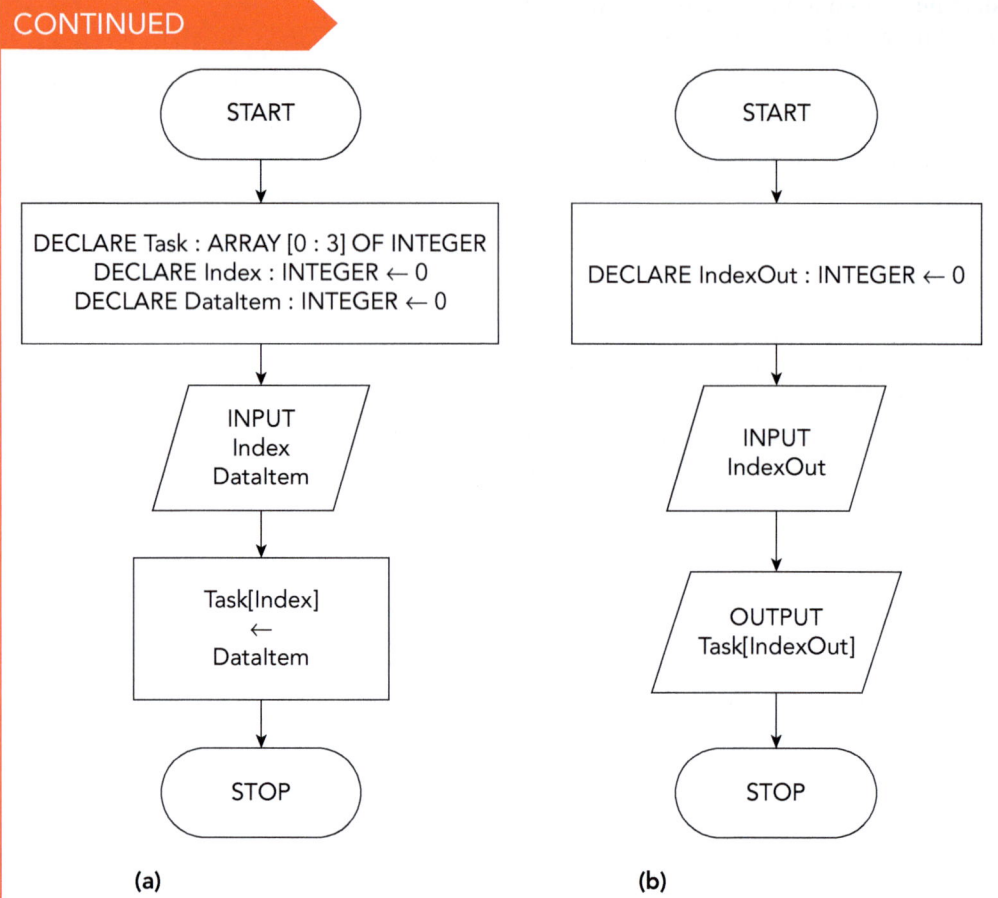

Flowchart 10.1: Flowcharts for the INPUT (a) and OUTPUT (b) process

Input pseudocode:

```
DECLARE Task : ARRAY [0 : 3]
  OF INTEGER
DECLARE Index : INTEGER ← 0
DECLARE DataItem : INTEGER ← 0
OUTPUT "Enter index and data
  value"
INPUT Index
INPUT DataItem
Task [Index} ← DataItem
```

Output pseudocode:

```
DECLARE IndexOut : INTEGER ← 0


OUTPUT "Enter index to show"


INPUT IndexOut


OUTPUT Task[IndexOut]
```

Step 2 Consider how the user will indicate which task they wish to complete. The flowchart (Flowchart 10.2) shows how this could be approached using a console application approach. (Windows Forms application is optional content. It is not covered in the syllabuses.)

> **CONTINUED**
>
> **Step 3** Consider how the user will indicate the array index required. These examples assume the user is familiar with the concept of 'zero-based arrays'. Had this not been the case, the input value would need to be reduced by 1 before use.

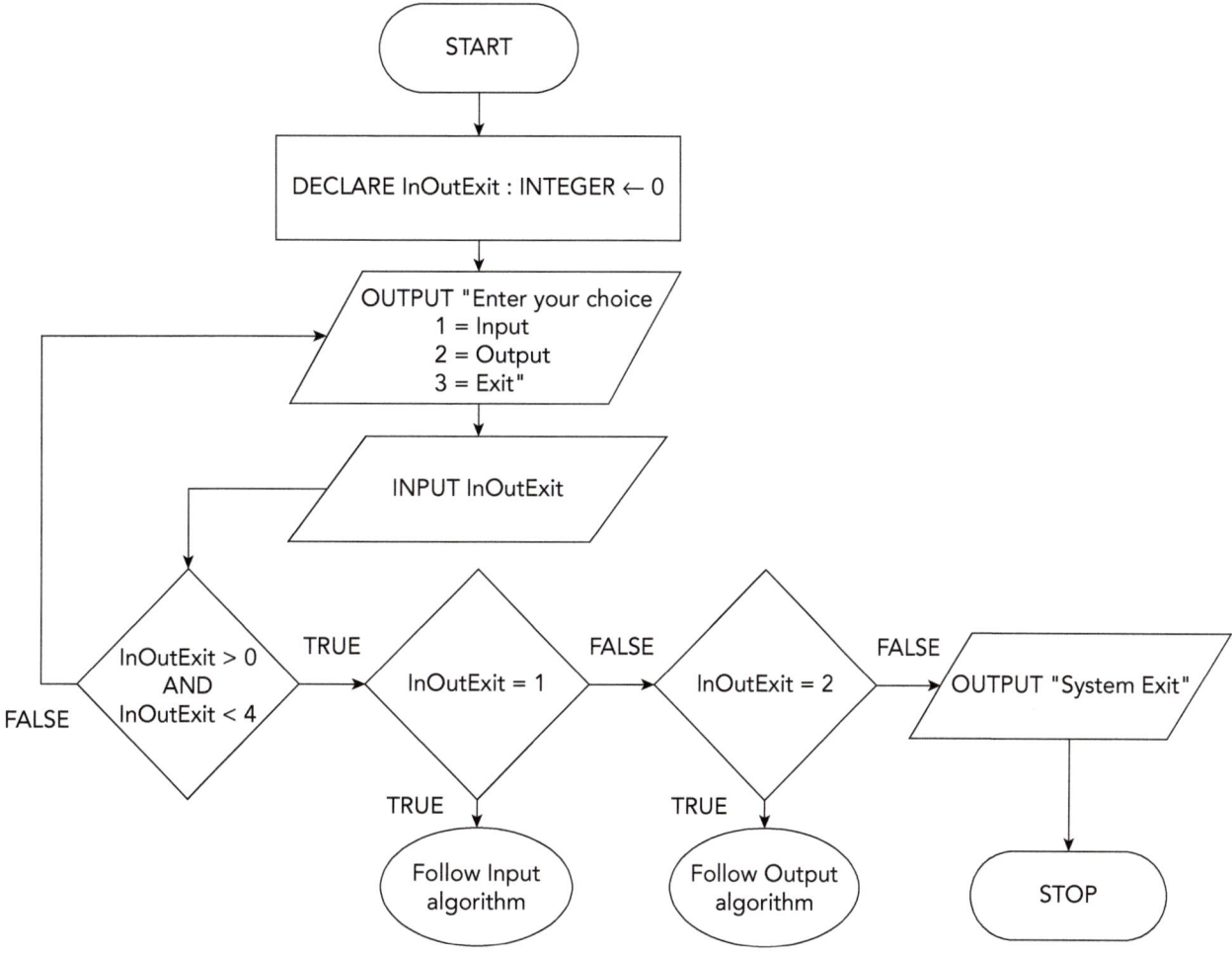

Flowchart 10.2: Flowchart to show the array index.
NOTE: Observe the use of the END symbol altered to indicate that the flow follows into another flowchart. It is not unusual to see this approach used in complex diagrams to avoid overly complex diagrams.

The flowchart also includes and initial decision to check the user input is within the expected range.

> **Note:** Windows Forms application is optional content. It is not covered in the syllabuses.

A Windows Forms application could be used to produce the system. The interface could be designed as shown in Figure 10.2 with a button to run the subroutines for each of the input and output processes. Textboxes accept the input and display the output.

> **CONTINUED**

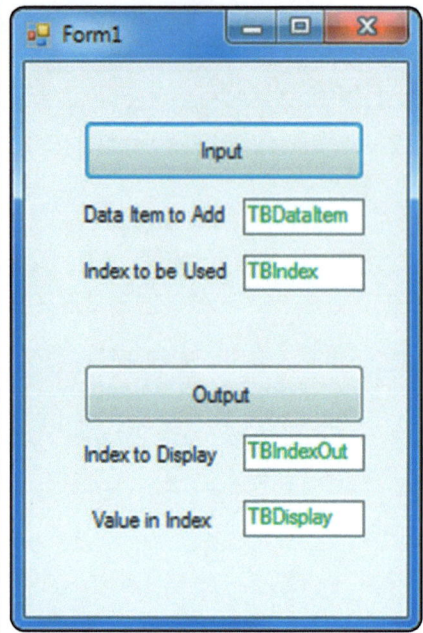

Figure 10.2: Windows Forms application

The names of the textboxes used in the example code are shown in green. The code required to produce the system would be as follows:

```vb
Public Class Form1
    'Global declaration of the array to give scope to both subroutines
    Dim Task(3) As Integer
    Private Sub BTNInput _ Click(sender As Object, e As EventArgs) Handles BTNInput.Click
        'Local declaration and initialisation of Input variables
        Dim Index As Integer = 0
        Dim DataItem As Integer = 0
        'Obtaining user input
        Index = TBIndex.Text
        DataItem = TBDataItem.Text
        'Input of DataItem into selected index
        Task(Index) = DataItem
    End Sub

    Private Sub BTNOutput _ Click(sender As Object, e As EventArgs) Handles BTNOutput.Click
        'Local declaration and initialisation of Output variable
        Dim IndexOut As Integer = 0
        'Obtaining user selection of index to display
        IndexOut = TBIndexOut.Text
        'Displaying the data item at the selected index in the Textbox
        TBDisplay.Text = Task(IndexOut)
    End Sub
End Class
```

CONTINUED

As an alternative, the same outcome could be achieved by using a console application with two subroutines. One subroutine for the input and the other for the output.

```vb
Module Module1
    'Global declaration of array to give scope to both subroutines
    Dim Task(3) As Integer


    Sub Main()
        'Declare variable to hold user choice
        Dim InOutExit As Integer
        'start Repeat..Until loop to validate input
        Do
            Console.WriteLine("Enter 1 to input, 2 to output, 3 to exit")
            InOutExit = Console.ReadLine()
        Loop Until InOutExit > 0 And InOutExit < 4


        'IF statement to call appropriate sub-routines
        If InOutExit = 1 Then
            ArrayIn()
        ElseIf InOutExit = 2 Then
            ArrayOut()
        Else
            Console.WriteLine("System Exit")
            Console.ReadKey()
        End If
    End Sub


    Sub ArrayIn()
        'Declare local variables to hold user input
        Dim Index As Integer
        Dim DataValue As Integer
        'Prompt for and store user input of index location and data value
        Console.WriteLine("Enter the index location")
        Index = Console.ReadLine()
        Console.WriteLine("Enter the input value")
        DataValue = Console.ReadLine()
        'Write DataValue to the array 'Task' at Index location
        Task(Index) = DataValue
        'Call Main sub-routine to allow user choice
        Main()
        Console.ReadKey()
    End Sub
```

> **CONTINUED**

```
    Sub ArrayOut()
        'Local declaration to hold user input
        Dim Index As Integer
        'Prompt for and store user input
        Console.WriteLine("Enter the index location")
        Index = Console.ReadLine()
        'Output value in the array 'Task' at Index location
        Console.WriteLine(Task(Index))
        'Call Main sub-routine to allow user choice
        Main()
        Console.ReadKey()
    End Sub
End Module
```

Question

Design the pseudocode and then create a system that stores up to 10 student names in an array called '`Student`'. The system will allow a user to input and output values as follows:

Input	The user will input the required index location and the name to be stored in that index location. The system will store the name at that location, overwriting any name already stored in that location.
	Validate the input to only accept valid index locations.
Output	The user will input the required index location. The system will output the value stored in that index location. If the index location indicated is empty the system will return an empty string (this is the default value for a string data type array).
	Validate the input to only accept valid index locations.

10.5 Iteration in arrays

The process of reading individual array positions can be extended by using iteration to read all the positions in an array. This allows iterative code (an iteration) to be used to check multiple data values.

Where the size of the array is known, a FOR loop can achieve the required iterative process. The counter variable in the FOR loop can be used to iterate through the index positions.

The following pseudocode shows how iteration can be used to either **a**, write to all array index locations or **b**, output all the data items in a ten-item string array called 'Test':

a Using a loop to write to array 'Test'. Will fill all index locations with the word 'Empty'.

```
DECLARE Test : ARRAY [0 : 9] OF STRING
    FOR Counter ← 0 TO 9
        Test [Counter] ← "Empty"
    NEXT Counter
```

b Using a loop to read from array 'Test'. Will output the value in all the index locations.

```
FOR Counter ← 0 TO 9
    OUTPUT Test [Counter]
NEXT Counter
```

10 Arrays

Linear search

Using a loop to read from an array can also be used to search the array. The code within the loop will search all the locations in the array. This is known as a linear search.

The search will start at the first index of the array and check every index location to see if the required data is present. This search will continue until the data required is found or the end of the array is reached.

> **KEY WORD**
>
> **linear search:** a method of searching an array. The process checks every element in the array sequentially until the required element is found or the end of the array is reached.

DEMO TASK 10.1

Array letters

Declare an array called 'Letters' to hold six single characters. Initialise the array with letters A to F. Write code that allows the user to input a letter and then search the array to identify if the letter input is in the array.

Solution

Flowchart 10.3 shows a flowchart for the Letters array.

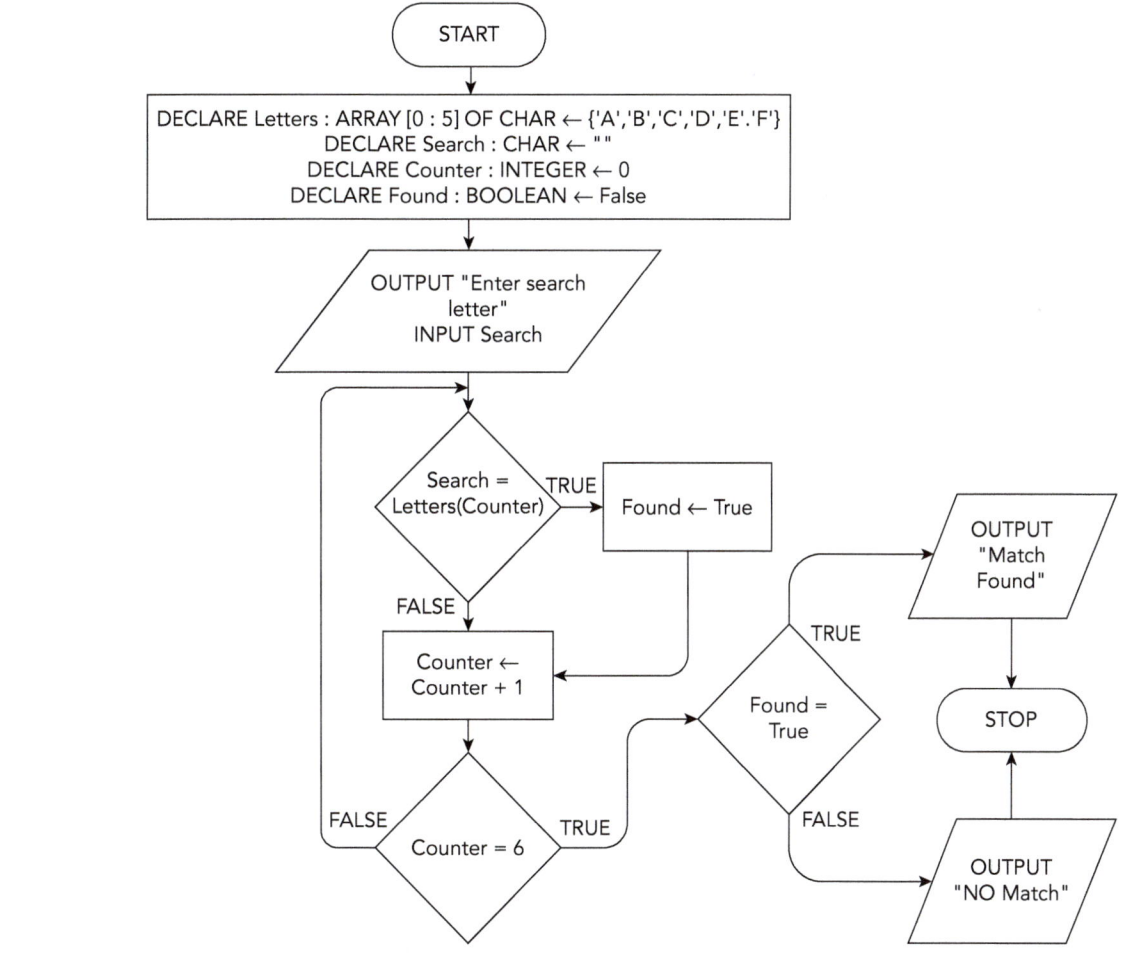

Flowchart 10.3: Flowchart for Letters array

> **CONTINUED**

```
    DECLARE Letters : ARRAY [0 : 5] OF CHAR ← {'A','B','C','D','E'.'F'}
    DECLARE Search : CHAR ← ""
    DECLARE Found : BOOLEAN ← False

    OUTPUT "Enter search letter"
    INPUT Search


    FOR Counter ← 0 TO 5
        IF Search = Letters[Counter]
           THEN
             Found ← True
        ENDIF
    NEXT Counter


    IF Found = True
      THEN
        OUTPUT "Match Found"
      ELSE
        OUTPUT "No Match"
    ENDIF
```

The following coded solution uses a console application:

```vb
Module Module1
    'Declare array - NOTE no size declared as array is initialised
    Dim Letters() As Char = {"A", "B", "C", "D", "E", "F"}


    Sub Main()
        Dim Search As Char = ""
        Dim Found As Boolean = False
        Console.WriteLine("Enter search letter")
        Search = Console.ReadLine()


        'Loop limit obtained by using 'length - 1' as array indexed from zero
        For Counter = 0 To Letters.Length - 1
            'Selection used to set boolean to true if match
            ' identified at any array location

            If Search = Letters(Counter) Then
                Found = True
            End If
        Next


        'Selection statement used to output appropriate message
        ' based on the value of the boolean variable found
```

CONTINUED

```
            If Found = True Then
                Console.WriteLine("Match Found")
            Else
                Console.WriteLine("No Match")
            End If
            Console.ReadKey()
        End Sub
End Module
```

PRACTICE TASKS 10.1–10.3

10.1 Student name array 2

Alter the Skills Focus 10.1 'student name' array:

a To allow the user to input a search value and then search the array to identify if a name held in the array matches. The system will output either 'Name Found' or 'No Match'.

b To allow the user to input a search value and then search the array to identify if a name held in the array matches. The system will output the index location of any matching name (for example, if the array held the name 'John' in two different index locations, a search for 'John' would output two values).

10.2 Letters array

Demo Task 10.1, 'Array Letters', is not very efficient. The loop will search all locations in the array, even if a match has been identified early in iteration. For an array with six locations, this will make little difference. However, for a large array, this inefficiency may cause the code to iterate many more times than required. Consider where an array used to hold 100 000 unique mobile phone numbers is searched. In the worse-case scenario, where the first index location holds the matching value, the FOR loop would continue to execute unnecessarily for a further 99 999 times.

Use a conditional loop so that the iteration will stop when the match is identified and output the message 'Match Found'. It will continue only to the final array index location in the event of no previous match, at which time, the message 'No Match' should be output.

10.3 Random numbers

It is possible to generate random numbers in Visual Basic. You can achieve this is two ways. Either use the Rnd() function described in Chapter 3. As an alternative you could create a local object of the inbuilt Random Class. This approach has the advantage that you can easily create an upper limit and lower limit for the range of random numbers. The following code will output 10 random numbers between 0 and 499. It creates a local object called 'Rand' and then uses the Next function to generate random numbers within a FOR loop. The two parameters passed to the Next function indicate the maximum range for the random numbers. The second number is exclusive and not included in the range.

> **CONTINUED**
>
> ```
> Sub Main()
>     Dim Rand As New Random
>     For Counter = 1 To 10
>         Console.WriteLine(Rand.Next(0, 500))
>     Next
>     Console.ReadKey()
> End Sub
> ```
>
> Using either console application or a Windows Forms application, write a system that will use the random number generator to initialise an array with 100 random numbers. The code should:
>
> 1 Output the total of all 100 numbers.
>
> 2 Output the highest and lowest number in the array.
>
> 3 Output the average of the numbers in the array.
>
> 4 Allow the user to input a number and the system will indicate if it is in the array. If the number is in the array the appropriate index location will be returned.

10.6 Grouped data records

While a single one-dimensional array can be used to hold a set of related, single-value data items, it is not able to easily hold more than one data element for each record.

One solution that can be used to hold multiple data elements for each data record is to use one-dimensional arrays in groups. Provided the same index number is used in each array for the data items that relate to one record, the values of multiple data items can be read.

Consider a situation where a system holds the records shown in Table 10.2:

Student ID	Surname	Computing Grade
1001	Rodriguez	A
1002	Ali	C
1003	Clarke	B

Table 10.2: Student records

These data items could be held in three arrays as shown in Tables 10.3, 10.4 and 10.5 with the same index position in each array referring to the data regarding one record. For example, ID 1002 is held at index position 1 in the one-dimensional array; therefore, the remaining data for that record is also held in index position 1 in the other two arrays.

Index	0	1	2
Data Item	1001	1002	1003

Table 10.3: `StudentID` array

Index	0	1	2
Data Item	Rodriguez	Ali	Clarke

Table 10.4: Surname array

Index	0	1	2
Data Item	A	C	B

Table 10.5: Grade array

The pseudocode to output the Surname and Grade for a given Student ID would be as follows:

```
OUTPUT "Enter StudentID search value"
INPUT SearchID


FOR Counter ← 0 to 2
    IF SearchID = StudentID(Counter)
      THEN
         OUTPUT Surname(Counter)
         OUTPUT Grade(Counter)
    ENDIF
NEXT Counter
```

The logic of this pseudocode is to search the `StudentID` array for a match. If a match is identified, then the same index location of the match is used to output the surname and grade. The index location is determined by the value of the loop counter.

DEMO TASK 10.2

Baby data

Design the pseudocode and create a system that could be used to hold and search the following data that records the details of children born at a local hospital.

BabyID	Gender	Weight (Kg)	Blood Group
B2003	Male	3.50	O Neg
B2004	Female	3.34	A Pos
B2005	Female	3.62	O Pos

It should be possible to search the system by `BabyID` outputting all the details of the child.

Step 1 The array will need to declare and initialise four individual one-dimensional arrays to hold the data.

Step 2 The system will need to obtain the `BabyID` to search.

Step 3 A loop is needed that will compare the `BabyID` input with all the IDs in the array that is holding the baby IDs.

Use iteration with the loop counter value being used to iterate through the index locations.

Loop counter will have to start at zero as first index is zero.

> **CONTINUED**
>
> Could use FOR loop or WHILE loop – as `BabyID` is unique, a WHILE loop that stops when a match is found would be more efficient than a FOR loop, which will always search all locations.
>
> **Step 4** If a match is found, will need to record the index location and then output the values in the other one-dimensional arrays at the same index location as the match.
>
> **Step 5** If no match is found, will need to output a message.
>
> Solution
>
> Flowchart 10.4 shows the flowchart solution.
>
>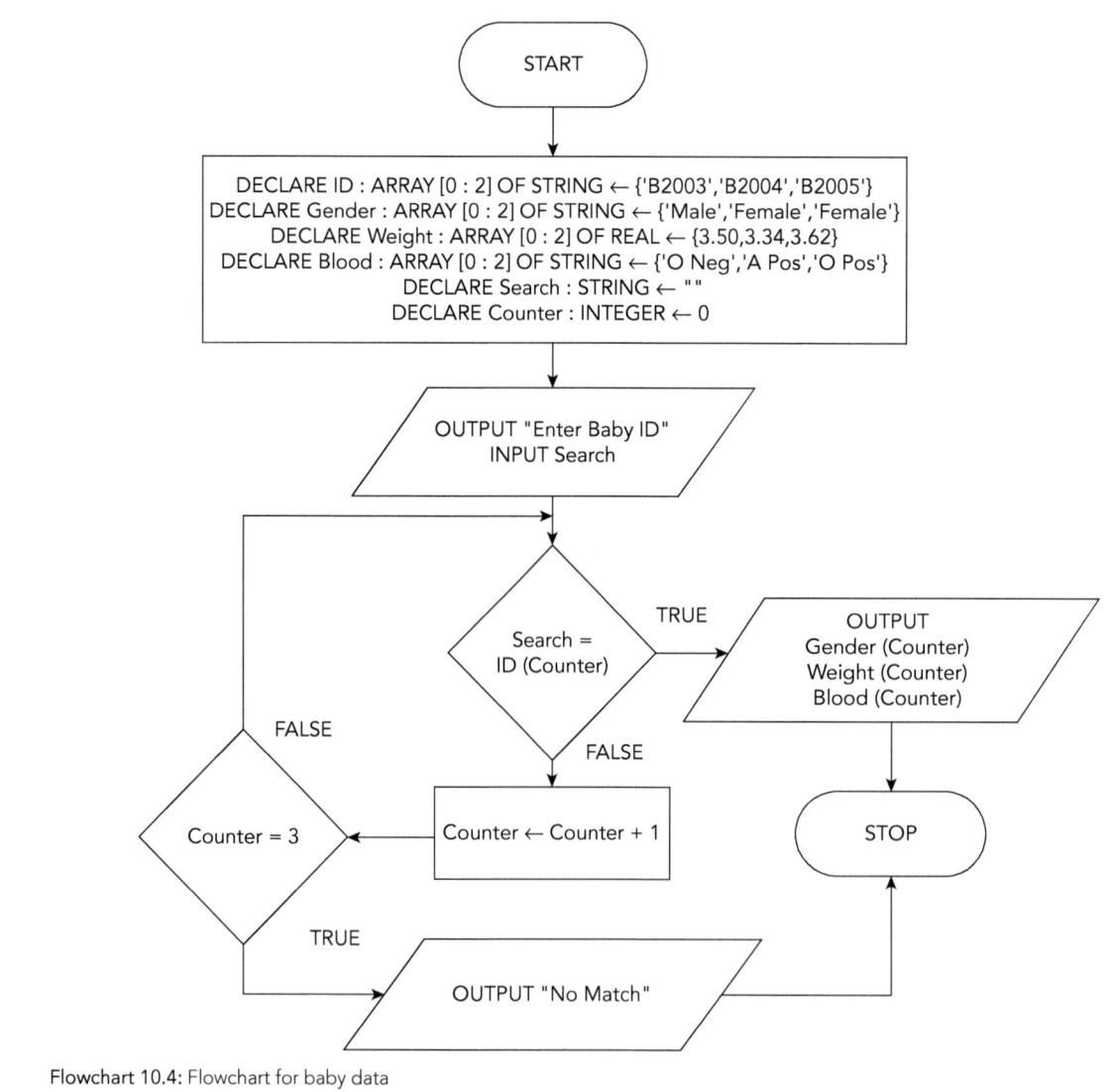
>
> Flowchart 10.4: Flowchart for baby data

CONTINUED

Here is the pseudocode solution:

```
DECLARE ID : ARRAY [0 : 2] OF STRING ← {'B2003','B2004','B2005'}
DECLARE Gender : ARRAY [0 : 2] OF STRING ← {'Male','Female','Female'}
DECLARE Weight : ARRAY [0 : 2] OF REAL ← {3.50,3.34,3.62}
DECLARE Blood : ARRAY [0 : 2] OF STRING ← {'O Neg','A Pos','O Pos'}
DECLARE Search : STRING ← ""
DECLARE Counter : INTEGER ← 0
DECLARE Found : BOOLEAN ← False

OUTPUT "Enter ID to search"
INPUT Search

WHILE Counter < 3 AND Found = False DO
    IF Search = ID [Counter]
       THEN
          OUTPUT Gender (Counter)
          OUTPUT Weight (Counter)
          OUTPUT Blood (Counter)
          Found = True
    ENDIF
    Counter ← Counter + 1
ENDWHILE
IF Found = False
   THEN
      OUTPUT "No Match Found"
ENDIF
```

```vb
Module Module1
    'Declare and initialise the required arrays
    'Ensure the date values are in the correct order
    Dim ID() As String = {"B2003", "B2004", "B2005"}
    Dim Gender() As String = {"Male", "Female", "Female"}
    Dim Weight() As Decimal = {3.5, 3.34, 3.62}
    Dim Blood() As String = {"O Neg", "A Pos", "O Pos"}

    Sub Main()
        'Dim local variables to use with search
        Dim Search As String = ""
        Dim Counter As Integer = 0
        Dim Found As Boolean = False
        'Prompt for and store user input
        Console.WriteLine("Enter ID to search")
        Search = Console.ReadLine()


        'While loop with dual criteria - both must be true to run
        While Counter < 3 And Found = False
            'Check for match against ID array
```

> **CONTINUED**

```
            If Search = ID(Counter) Then
                'Output the corresponding values from other arrays
                Console.WriteLine(Gender(Counter))
                Console.WriteLine(Weight(Counter))
                Console.WriteLine(Blood(Counter))
                'Set found to true to end loop
                Found = True
            End If
            'increment counter to search next index location
            Counter = Counter + 1
        End While


        'If found still false - no match found so output message
        If Found = False Then
            Console.WriteLine("No match found")
        End If


        Console.ReadKey()
    End Sub
End Module
```

PRACTICE TASK 10.4

Baby data

Alter the code in Demo Task 10.2 so that it is possible to:

a   Search by Gender – outputting the `BabyID` of all records that match the gender type.

b   Output the average weight of all the babies in the array.

TIP

Loop through the array to add the weight of all babies and divide that answer by the number of babies.

CHALLENGE TASK 10.1

Design a record system for you and your colleagues. It could be based on any shared interest such as a shared interest in sports. It would be useful if a unique identifier could be used for each record. An example could be:

`StudentID`	First Name	Favourite Sport	Play for a Team	Hours Played

> **CONTINUED**
>
> Decide how many records your system will keep and create appropriate arrays for each data item.
>
> The system will need to have an input facility so that the details can be input. Consider how you might avoid overwriting existing details when adding new data. One approach is to search for the first blank index and then adding new data to that index.
>
> Use the array to provide appropriate search facilities. In the example shown, you could search for:
>
> - All people who have the same favourite sport.
> - How many people play for a team.
> - The average amount of hours played.
> - The maximum number of hours played.

10.7 Two-dimensional arrays

An alternative approach to using multiple one-dimensional arrays to hold grouped data is to use a single **two-dimensional array**.

Many people visualise a two-dimensional array as a grid, indexed in both the horizontal and vertical dimensions. Each index location being identified by using both index locations as shown in Table 10.6:

Index	0	1	2
0	Location (0,0)		
1			
2		Location (2,1)	Location (2,2)

Table 10.6: Locations in a two-dimensional array

> **KEY WORD**
>
> **two-dimensional array:** an array with two index sets. Contains multiple rows of data with multiple columns. Each individual element identified by a combination of both row and column index.

Table 10.7 is a visual representation of a two-dimensional array named `Student` used to hold `StudentID`, Surname and Exam Grade data for four computer science students.

Index	0	1	2
0	1001	Rodriguez	A
1	1002	Ali	C
2	1003	Clarke	A
3	1004	Devi	B

Table 10.7: Visual representation of a two-dimensional array

The pseudocode for declaring this two-dimensional array is:

```
DECLARE Student: ARRAY [0:3, 0:2] OF STRING
```

The pseudocode to output values in a two-dimensional array is:

`OUTPUT Student [0,0]` would output the value '1001'
`OUTPUT Student [2,1]` would output the value 'Clarke'
`OUTPUT Student [3,1]` would output the value 'Devi'

The syntax for declaring the two-dimensional array `Student` in Visual Basic is shown in Figure 10.3.

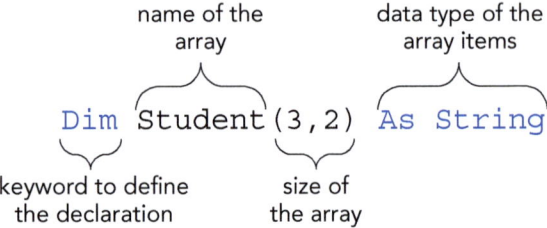

Figure 10.3: Visual Basic declaration syntax for a two-dimensional array

In Visual Basic, the size of the array is defined by stating the final index number of both indexes. The first index position of zero is assumed.

The order of the indexes is unimportant. While we may find it helpful to visualise an array as a grid, the array is not stored as a grid in your computer memory. Therefore the concept of rows and columns is used only to help programmers visualise an array.

Two approaches that could be used when declaring an array to hold identical data are shown in Table 10.8:

| `Dim Student (3,2) As String` | `Dim Student (2, 3) As String` |

Table 10.8: Comparisons of two orientations of data

This could be visualised as similar grids with different orientation:

Index	0	1	2
0	1001	Rodriguez	A
1	1002	Ali	C
2	1003	Clarke	A
3	1004	Andrews	B

Index	0	1	2	3
0	1001	1002	1003	1004
1	Rodriguez	Ali	Clarke	Devi
2	A	C	A	B

When using the arrays, it is crucial to match the method of declaration with the related code.

| To output the name Clarke would require:

`OUTPUT Student (2,1)` | To output the name Clarke would require:

`OUTPUT Student (1,2)` |

DEMO TASK 10.3

Create a system that uses a two-dimensional array called 'Employee' that could be used to hold and search the following records of employees at a delivery business.

EmloyeeID	Surname	Driver?	Salary
E104D	Wang	Yes	£25000
S135D	Abdou	Yes	£30000
J214	Ivanova	No	£26000

It should be possible to search the system by `EmployeeID`, outputting all the details of the employee.

Step 1 The array will need to be declared and initialised with the data values.

It is important that selection of indexes is considered. To match the visualisation of the data table, this example has chosen to represent as three rows and four columns. The three rows indicated by the first index and the four columns represented by the second.

This will make the index declaration in pseudocode [0:2 , 0:3].

NOTE: Although in pseudocode, both the start an end limits of each index are stated, this is not the same as an array declaration in Visual Basic. In Visual Basic, the array is either declared using only the end limit – `Dim Employee(2, 3) As String` – or through the initialisation process (see Code snippets 10.1 and 10.2).

Step 2 The system will need to obtain the `EmployeeID` that is to be searched.

Step 3 As the search is to be carried out on the `EmployeeID`, the code will need to iterate through all the array elements that hold the `EmployeeID`. These are [0:0], [1:0] and [2:0].

It can be seen that the first index value (the row) is incrementing, but the second index (the column) remains at 0. Therefore, the loop counter will need to impact on the first array index.

Step 4 If a match is found in a particular row, will need to output all the other values for that record by incrementing through the column index locations for that row.

Step 5 If no match is found, will need to output message. Use a Boolean to indicate if a match is found. Change to TRUE only when a match is found. After the loop has executed, check value of it remains at FALSE then no match was found so output message.

CONTINUED

Solution

Flowchart 10.5 shows the flowchart solution.

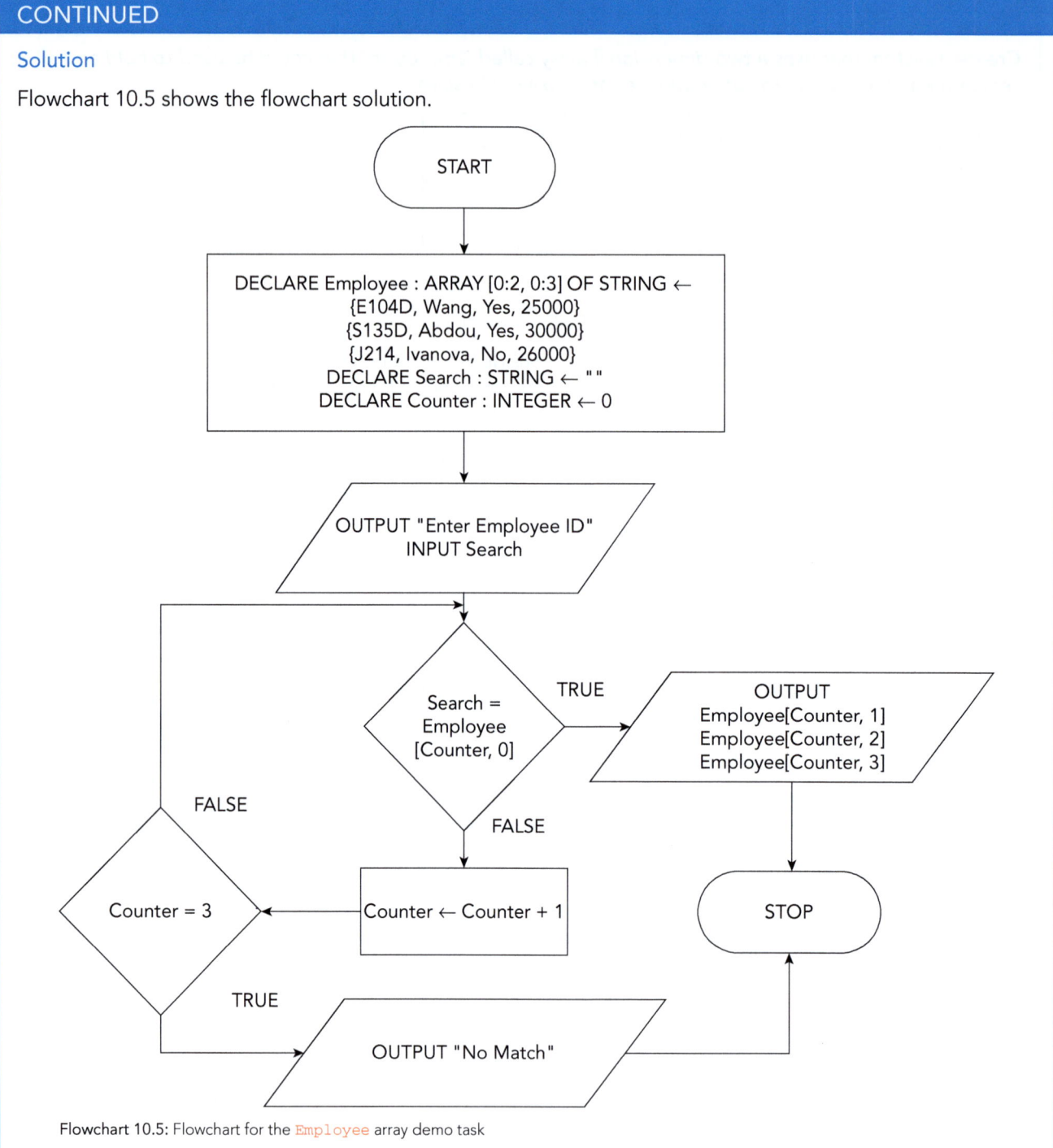

Flowchart 10.5: Flowchart for the Employee array demo task

CONTINUED

Here is the pseudocode solution:

```
DECLARE Employee : ARRAY [0:2, 0:3] OF STRING ← {E104D, Wang, Yes, 25000}
{S135D, Abdou, Yes, 30000} {J214, Ivanova, No, 26000}
DECLARE Search : STRING ← ""
DECLARE Counter : INTEGER ← 0
DECLARE Found : BOOLEAN ← False


OUTPUT "Enter Employee ID"
INPUT Search


WHILE Counter < 3 AND Found = False DO
    IF Search DO = Employee[Counter, 0]
      THEN
        OUTPUT Employee [Counter, 1]
        OUTPUT Employee [Counter, 2]
        OUTPUT Employee [Counter, 3[
        Found ← True
    ENDIF
ENDWHILE
IF Found = False
  THEN
    OUTPUT "No Match"
ENDIF


IF Found = False
  THEN
    OUTPUT "No Match"
ENDIF
Module Module1
    'Declare and initialise the array
    'NOTE array dimensions not input as size defined by initialisation
      'the array is initialised as if it is a multiple 1D arrays
      'each { } defines a new 'row' in the array
    Dim Employee(,) As String = {{"E104D", "Wang", "Yes", "25000"},
      {"S135D", "Abdou", "Yes", "30000"}, {"J214", "Ivanova", "No", "26000"}}
    Dim Search As String = ""

    Dim Counter As Integer = 0
    Dim Found As Boolean = False


    Sub Main()
        Console.WriteLine("Enter Employee ID")
        Search = Console.ReadLine()
        'Set while loop with dual criteria
```

> **CONTINUED**
>
> ```vb
>             Do While Counter < 3 And Found = False
>                 'Use Counter to loop through first index of array
>                 'As the ID is held in first element of the second index
>                 ' the index used is (Counter, 0) as this will loop through all
>                 'rows'
>                 ' but always use the first (0 index) column
>                 If Search = Employee(Counter, 0) Then
>                     'Output the values in the remaining column indexes
>                     ' at the same Counter value as the match
>                     Console.WriteLine(Employee(Counter, 1))
>                     Console.WriteLine(Employee(Counter, 2))
>                     Console.WriteLine(Employee(Counter, 3))
>                     'Set found to true to stop loop as EmployeeID is unique
>                     Found = True
>                 End If
>             Loop
>             'IF after no value in array matches with the input employee ID
>             ' Found will still be False and 'No Match' message is output
>             If Found = False Then
>                 Console.WriteLine("No Match")
>             End If
>
>
>             Console.ReadKey()
>         End Sub
>     End Module
> ```
>
> Code snippet 10.1
>
> Some programmers find the initialising syntax of a two-dimensional array unnecessarily complex and prefer to use a different approach. They declare the array including the index dimensions and then input the values into the array as part of the execution code:
>
> ```vb
>     Dim Employee(2, 3) As String
>
>         Sub Main()
>             Employee(0, 0) = "E104D"
>             Employee(0, 1) = "Wang"
>             Employee(0, 2) = "Yes"
>             Employee(0, 3) = "25000"
>             Employee(1, 0) = "S135D"
>             Employee(1, 1) = "Abdou"
>             Employee(1, 2) = "Yes"
>             Employee(1, 3) = "30000"
>             Employee(2, 0) = "J214"
>             Employee(2, 1) = "Ivanova"
>             Employee(2, 2) = "No"
>             Employee(2, 3) = "26000"
> ```
>
> Code snippet 10.2

When to use a two-dimensional array

In the examples given so far, using either multiple one-dimensional arrays or a single two-dimensional array would have been equally successful. In these situations, the choice is therefore dependent on the preference of the programmer.

Becoming familiar with the use of two-dimensional arrays could prove to be a real advantage as you progress on to more complex programming and meet situations that are uniquely suited to the use of a two-dimensional array.

Consider the situation where an array is being used to hold data normally structured in a grid format.

A traditional board game such as Chess, where the playing pieces can move in all directions.	Word games such as a crossword or word search where the letters representing one word can be written any direction.
If stored as a series of eight individual one-dimensional arrays, recording a diagonal forward move would involve several different arrays.	If stored as a series of individual one-dimensional arrays, to store the word 'wordsearch', in this example, would involve nine different arrays.

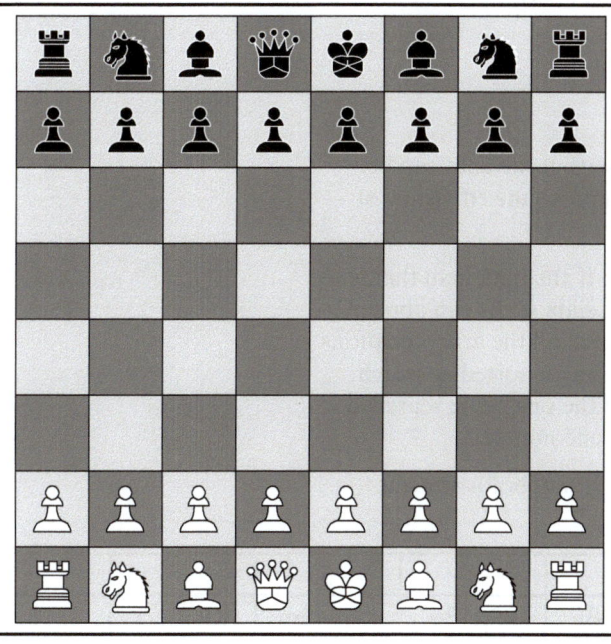 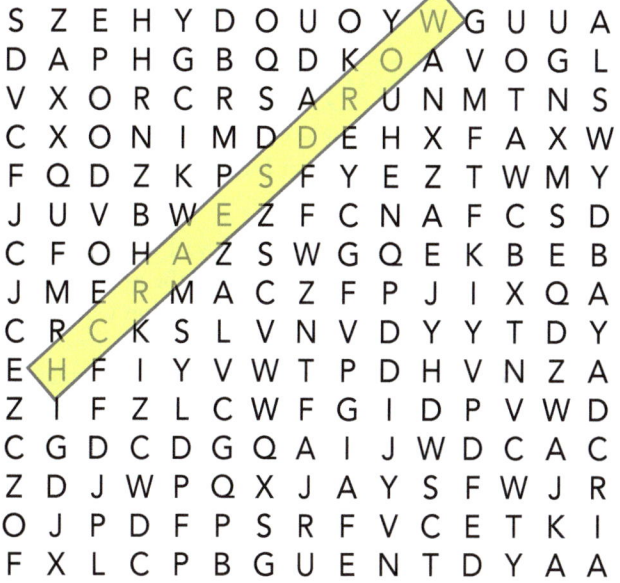

In these examples, a single two-dimensional array that allows code to iterate through either of the indexes has an advantage over a series of one-dimensional arrays. For example, when storing a word into the electronic word search array, a loop could be used to increment either one or both index locations to place each character of the word next to each other in the array. The way in which the indexes were incremented could determine the direction of the word. Incrementing one index would result in a vertical or horizontal placement of letters. Incrementing both indexes would result in diagonal placement of letters.

Another advantage of using two-dimensional arrays is that all the data in the array can be read or written by a combination of two loops. Consider a two-dimensional array `Chess` that has been created to store the positions of the chess pieces in an electronic chess game. The individual pieces are stored as string values, with the white queen stored as WQ. To search for the location of the white queen, the following double loop could be used:

```
Dim Chess(7, 7) As String
Dim Row As Integer = 0
Dim Column As Integer = 0


For RowCounter = 0 To 7
    For ColumnCounter = 0 To 7
        If Chess(RowCounter, ColumnCounter) = "WQ" Then
            Row = RowCounter
            Column = ColumnCounter
        End If
    Next
Next
```

The logic of this approach is that for every iteration of the 'RowCounter' loop, the internal 'ColumnCounter' loop will iterate 8 times. If you complete a trace table, you will see how this dual loop will search every location in the array.

10.8 Sorting

Sorting the data in an array is a common task. This is done so that the data can be output in an order that is easier for a human to read, or to improve the efficiency of a search algorithm.

Sorting data will improve the efficiency of a search algorithm if the data item that is being searched for is not in the array. If the array is unsorted, data items can appear anywhere in the array, and a search algorithm will need to check all the array locations to confirm that the searched value is not in the array. If the array is sorted, a search algorithm can stop the search as soon as a value greater than the one being searched for (a bigger number or a letter further on in the alphabet) is encountered.

Consider the following alphabetically sorted array (Table 10.9) that holds student usernames:

Index	0	1	2	3	4	5	6	7
Data	AB145j	CA235f	CD764e	ES788f	FR218g	KH842m	KR156d	SF952g

Table 10.9: Student usernames array

A linear search is made for the username DR274h. The algorithm will start at index 0 and check the value of each of the array index locations to determine if they match the search value. The algorithm would be able to end the search when is checks index 3. The value at index location 3, ES788f, is alphabetically higher than the search value of DR274h. This means that the search value is not in the array because all the index locations above index 3 must hold alphabetically higher values than the search value.

Sorted arrays can be particularly efficient when a large number of data items are involved. Consider a search for the name Aaron in an alphabetically sorted array of 10 000 names. If Aaron is not in the array, this could be determined very quickly as the name would appear early in the array. This could save unnecessary checking the value in many of the index locations.

> **TIP**
>
> Remember that sorting the array will take time and computing power. The advantage of using a sorted array should be considered against the time and effort required to sort the array.

10.9 Sorting algorithms

There are many different sorting algorithms that have been created for programmers to use. Visual Basic in common with other languages includes a sort procedure: `Array.Sort('Array to sorted')` will sort both a numerical and string array. The default sort order is shown in Table 10.10:

Date type of array	Sort order
String	Alphabetical in ascending order (a to z).
	For words with the same first letter the sort will compare subsequent letters.
	A mix of upper case and lower-case letters will both be sorted in ascending order. Lowercase letters are considered lower than the equivalent upper-case letter.
	For example, 'alpha' is lower than 'Alpha' but both are lower than 'beta'.
Numerical (Integer, Decimal)	Numerical in ascending order.

Table 10.10: Sort order for `Array.Sort`

If you wish to sort the array in descending order, first use the `Array.Sort` procedure to sort the array then use the `Array.Reverse` procedure to reverse the order of the sorted array.

Thie following shows the code and the console application output (Figure 10.4) of the `Array.Sort` and `Array.Reverse` procedures:

```
Sub Main()
    Dim Letters() As String = {"z", "a", "zA", "Z", "f",
    "g", "aa", "ab", "Ab", "c"}
    Console.WriteLine("The array sorted ascending")
    Array.Sort(Letters)
    For Counter = 0 To 9
        Console.WriteLine(Letters(Counter))
    Next

    Array.Reverse(Letters)
    Console.WriteLine("The array sorted descending")
    For Counter = 0 To 9
        Console.WriteLine(Letters(Counter))
    Next
    Console.ReadKey()
End Sub
```

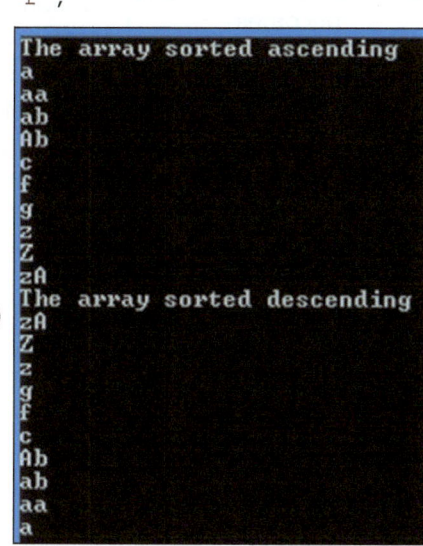

Figure 10.4: Console application output of the `Array.Sort` and `Array.Reverse` procedures

10.10 Bubble sort

While most languages include sort procedures, the way in which the procedures work is hidden from the programmer. Clearly an algorithm is called by the procedure, but the programmer does not know the detail of that algorithm. One common sorting algorithm is the **bubble sort** algorithm, and it is not unusual for teachers to expect students to know how a bubble sort algorithm works.

> **KEY WORD**
>
> **bubble sort:** an algorithm that is used to sort the values in an array into ascending or descending order.

The bubble sort algorithm sorts an array by comparing the values in consecutive index locations in the array. At each comparison, the higher value is moved towards the end of the array. Repeating this process will result in the highest value being at the end of the array and the lowest value at the beginning of the array, with the values in between in ascending order.

Consider this unsorted array of integer values:

Index	0	1	2	3
Data	7	11	2	4

The way a bubble sort algorithm would sort the data in this array in to ascending order is shown in Table 10.11. As the array holds four data items (last index location is 3) the algorithm will need to repeat the comparison process three times.

0	1	2	3	
				Iteration 1
7	11	2	4	Iteration 1 – Step 1. The values at index 0 and 1 are compared. As they are in the correct order no change is made.
7	11	2	4	Iteration 1 – Step 2. The values at index 1 and 2 are compared. As they are in the wrong order, the values are swapped.
becomes				
7	2	11	4	
7	2	11	4	Iteration 1 – Step 3 The values at index 2 and 3 are compared. As they are in the wrong order, the values are swapped. The last comparison of this iteration is complete and the value 11 is in the correct location.
becomes				
7	2	4	11	
				Iteration 2
7	2	4	11	Iteration 2 – Step 1. The values at index 0 and 1 are compared. As they are in the wrong order the values are swapped.
becomes				
2	7	4	11	
2	7	4	11	Iteration 2 – Step 2. The values at index 1 and 2 are compared. As they are in the wrong order the values are swapped. No need to compare with index location 3 as this is correct following iteration 1. This iteration is complete and the value 7 is in correct location.
becomes				
2	4	7	11	
				Iteration 3
2	4	7	11	Iteration 3 – Step 1. The values at index 0 and 1 are compared. As they are in the correct order no change is made. No need to compare with index location 2 as this is correct following iteration 2. This iteration is complete and the value 4 is in correct location. As no swaps have been made during this iteration all values must be in order and the algorithm can exit as the array is sorted.

Table 10.11: Bubble sorting integer values

> **PRACTICE TASK 10.5**
>
> This is a string array `Cities` which holds names of six capital cities.
>
> | London | Dhaka | New Delhi | Abu Dhabi | Washington DC | Port Louis |
> |---|---|---|---|---|---|
>
> Generate a trace table to show the steps a bubble sort would take to sort the array `Cities`.

Bubble sort algorithm

You may have noticed from the numbers example (Table 10.11) that a bubble sort consists of two iterative processes (see Table 10.12).

Loop	Explanation
Iteration loop	Controls the number of times the comparison loop will iterate. Should exit when no swaps have been made during one execution of the comparison loop. If no values are swapped the array must be sorted.
Comparison loop	Iterates through consecutive pairs of index numbers. Within the loop, the data values held at the pair of index locations are compared. If the data is not in the correct order, the data values are swapped. The loop will start iterate from two fewer than the number of items in the array. At each iteration the limit of the loop will be reduced by one as an increasing number of values become sorted removing the need for those values to be compared.

Table 10.12: Iterative processes of bubble sort

The bubble sort algorithm follows this logic by using nested iteration implemented by the use of two loops. The iteration loop provided by a conditional WHILE or UNTIL loop and the comparison loop provided by a FOR loop.

The flowchart of the bubble sort algorithm is shown in Flowchart 10.6

> **TIP**
>
> If the size of the array being sorted is not known, use the `array.length` command to return the number of index locations in the array.

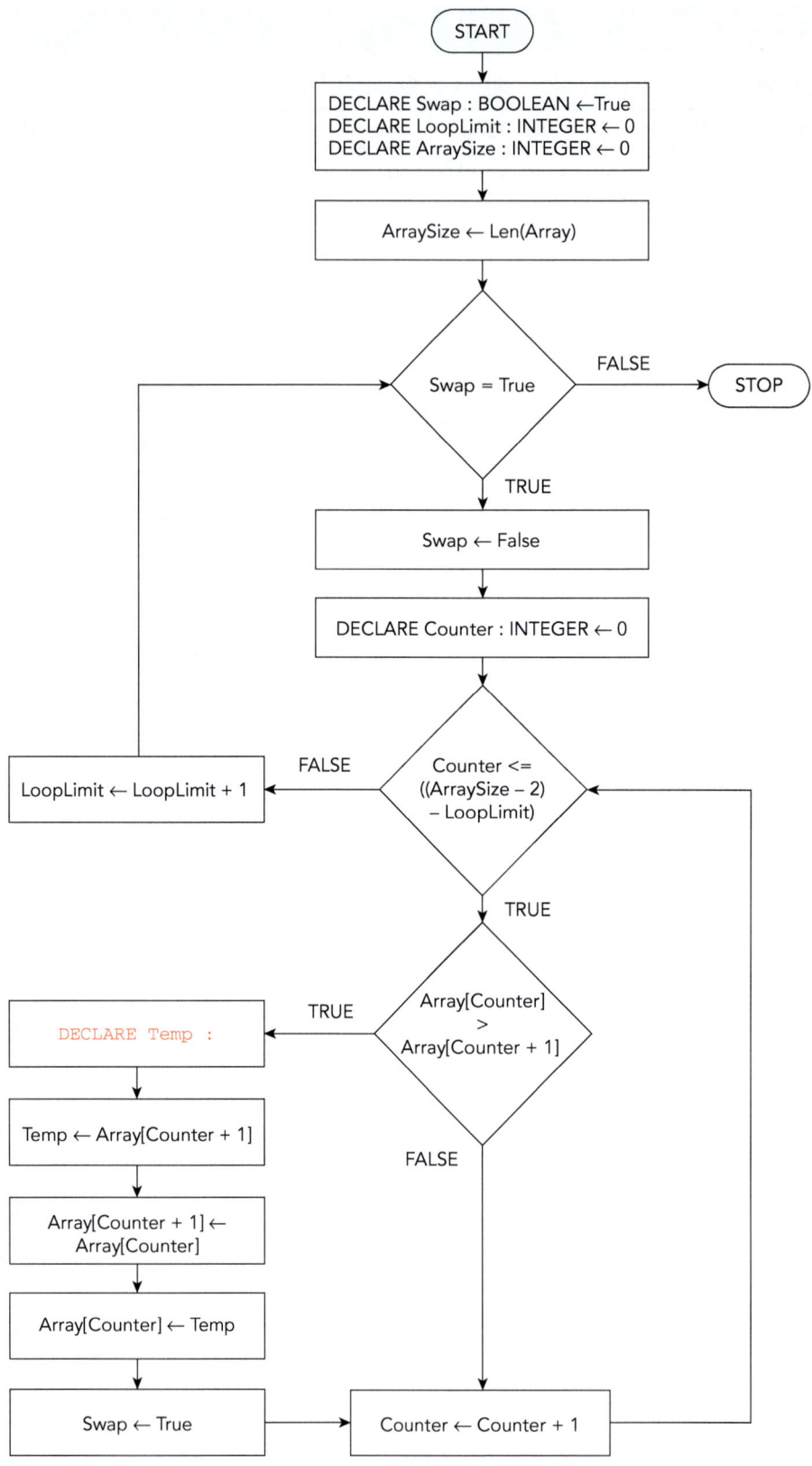

Flowchart 10.6: Flowchart for bubble sort algorithm

NOTE: the declaration of Temp shown in red, which has no datatype. This is because the datatype of Temp will depend on the datatype of the array being sorted. As Temp will be holding values from the array the datatypes must match to avoid datatype mismatch errors.

The pseudocode for a bubble sort algorithm can implement the conditional loop using either a WHILE or UNTIL loop. The following pseudocode shows the WHILE version of the bubble sort.

```
DECLARE Swap : BOOLEAN ← True
DECLARE LoopLImit : INTEGER ← 0
DECLARE ArraySize : INTEGER ← 0
//obtain the size of the array
ArraySize ← Len(Array)
WHILE Swap = True DO
    // Set Swap to false so if no swaps encountered the loop will exit
    Swap ← False
    //limit of loop will 2 less than the size of the array for zero indexed
    arrays
    //Looplimit will reduce the upper limit of the loop each time the loop is
    executed
    FOR Counter ← 0 TO ((ArraySize  - 2)  - LoopLimit)
        IF Array[Counter] > Array[Counter + 1]
          THEN
            //Complete the swap process
              DECLARE Temp : datatype to match the datatype of the array
              Temp ← Array[Counter + 1]
              Array[Counter + 1] ← Array[Counter]
              Array[Counter] ← Temp
              //as a swap has been completed set Swap to True to cause the WHILE
              loop to
                 run
              Swap ← True
        ENDIF
        // increment LoopLimit to reduce the end condition of the FOR loop as an
        //increasing number of values are in correct sorted location
    LoopLimit ← LoopLimit + 1
ENDWHILE
```

The REPEAT...UNTIL will check the value of Swaps in a different way. It will continue to loop until the value of Swap is False. This requires only two changes from the WHILE version which have been highlighted.

The following pseudocode shows the UNTIL version of the bubble sort.

```
DECLARE Swap : BOOLEAN ← True
DECLARE LoopLImit : INTEGER ← 0
DECLARE ArraySize : INTEGER ← 0
//obtain the size of the array
ArraySize ← Len(Array)
REPEAT
    // Set Swap to false so if no swaps encountered the loop will exit
  Swap ← False
    //limit of loop will 2 less than the size of the array for zero indexed
    arrays
    //Looplimit will reduce the upper limit of the loop each time the loop is
    executed
```

```
        FOR Counter ← 0 TO ((ArraySize  - 2)  - LoopLimit)
            IF Array[Counter] > Array[Counter + 1]
              THEN
                //Complete the swap process
                  DECLARE Temp : datatype to match the datatype of the array
                    Temp ← Array[Counter + 1]
                    Array[Counter + 1] ← Array[Counter]
                    Array[Counter] ← Temp
                    //as a swap has been completed set Swap to True to cause the UNTIL
                    loop to run
                    Swap ← True
            ENDIF
            // increment LoopLimit to reduce the end condition of the FOR loop as an
            //increasing number of values are in correct sorted location
            LoopLimit ← LoopLimit + 1
UNTIL Swap = False 230
```

The Visual Basic code for a bubble sort algorithm using a WHILE loop is:

```vb
Sub Main()
    Dim Numbers() As Integer = {52, 26, 3, 17, 46, 11, 18, 22, 2, 31}
    Dim ArraySize As Integer = 0
    'Obtain the size of the array that is being sorted
    ArraySize = Numbers.Length
    'Declare the Boolean variable to record if swaps are made
    'Set this to TRUE os the assumption that swaps will be made
    Dim Swaps As Boolean = True
    'Declare the variable that will be used to limit the
    range of the
    'FOR loop being used to control the comparison
    iteration
    Dim Counter2Limit As Integer = 0


    'WHILE loop to control the number of iterations of the
    compare loop
    While Swaps = True
    'Set the Boolean variable to FALSE as no swaps may be
    made this iteration
        Swaps = False
        'Compare loop to iterate through consecutive pairs
        of index locations
        'The variable Counter2Limit will reduce the maximum
        range at each iteration
        For Counter2 = 0 To ((ArraySize - 2) - Counter2Limit)
            'If statement compares values in the index locations
            'Using Counter2 and Counter2 + 1 as index numbers means
            'the comparison will always be between consecutive index locations
            'as the value of Counter2 increments different pairs are compared.
            If Numbers(Counter2) > Numbers(Counter2 + 1) Then
                Dim Temp As Integer = 0
                Temp = Numbers(Counter2)
                Numbers(Counter2) = Numbers(Counter2 + 1)
```

```
                Numbers(Counter2 + 1) = Temp
                'Set Boolean variable to TRUE as a swap has been made
                'This will cause the WHILE loop to run at least one more time
                Swaps = True
            End If
        Next
        'As this comparison loop is complete, increment the variable used to limit
        'the maximum range of the loop. This means the loop will end one value
        'earlier at each iteration.
        Counter2Limit = Counter2Limit + 1
    End While


    'Code to display the now sorted array
    For Counter = 0 To 9
        Console.WriteLine(Numbers(Counter))
    Next
    Console.ReadKey()
End Sub
```

PRACTICE TASK 10.6

Code a bubble sort algorithm that will sort an array into descending order.
NOTE: You only need to change one symbol.

SUMMARY

An array is a variable that can hold a set of data items of the same data type, under a single identifier.
When an array is declared, its size is defined. Visual Basic numbers elements in an array from zero. The size of an array can either be defined as part of the declaration or through the initialisation process.
Each element or data item in an array can be referenced by its index. The index can be used to read or write values in an array.
Arrays can be either one-dimensional or two-dimensional: • A one-dimensional array is a single row of data. • A two-dimensional array contains multiple rows of data.
A two-dimensional array is particularly useful when representing data normally represented in a grid format.
A one-dimensional array can be sorted into ascending or descending order. Sorted arrays can be more efficient to search.
A FOR loop can be used to iterate through the index locations in an array. The loop counter is used to identify successive index numbers.
A WHILE loop can be used to search index locations in an array. The loop can be exited once a match to the search value is found.

CONTINUED

Holding records which consist of more than one data item can be achieved by the use of multiple one-dimensional arrays. Data for each record is held at the same index position in the different arrays. As an alternative a two-dimensional array could be used with each row holding a different record.

Arrays can be sorted using either the inbuilt sort command or by running a bubble sort algorithm. Arrays holding ether string or numerical data types can be sorted.

END-OF-CHAPTER TASKS

The amount of data items in these tasks are deliberately small to allow you to code and test your system. While these tasks make use of limited data, the algorithms required to complete the tasks would work equally well with much larger data sets. The only change would be the number of iterations required as these are directly related to the number of data items.

1 An array Numbers is used to hold the following integer values:

17	103	36	42	85	3	64	98	55	11

  a Using pseudocode create a system that will use an array called Numbers to hold this data.

  b The system will use the array to output the following values:

   i   The sum of the numbers in the array.

   ii The largest and smallest numbers in the array.

   iii The mean average of the numbers in the array.

  c The system will have an option for the user to change individual numbers in the array.

  d Explain how your code would need to be altered if the array held 1000 numbers.

2 An array is to be used to identify if a six-digit number is a palindrome. A palindrome reads the same both forwards and backwards: 192291, 788887, 123321, 999999, 114411 are all examples of palindromes.

  The user will enter each of the six numbers in turn. Each number will be stored to an index location

  For example, if the number 786678 was entered it would be stored as follows:

Index	0	1	2	3	4	5
Number	7	8	6	6	7	8

  Once the number has been entered, the system will identify if the number is a palindrome and output an appropriate message.

  Design the pseudocode for an appropriate solution to the task. Program the system and run tests to make sure the code works as expected.

3 Pseudocode and create a system that will use a series of one-dimensional arrays to hold the following data about animals for sale.

Type	Colour	Suggested minimum age of children in household	Breed
Cat	White	5	Persian
Dog	Black and White	12	Husky
Dog	Grey	1	Poodle
Parrot	Blue and Red	8	Macaw

CONTINUED

The system should be capable of being searched as follows:

a The user inputs an animal type, and the system outputs all the data about animals of that type.

b The user inputs the age of the youngest child in the household, the system will output the data about suitable animals based on the minimum age suggestion.

c The user inputs both type and the age of youngest child in the household, the system will output the data about suitable animals based on the type wanted and the minimum age suggestion.
NOTE: One method to achieve the age restriction could be to ask the user to input 18 if they either have no children or they wish to ignore the suggested minimum child age. This would mean an age would always be available to compare in the array.

4 Reproduce Task 3 using a single two-dimensional array.

5 It is possible to generate random numbers in Visual Basic. To achieve this, you need to create a local object of the inbuilt `Random Class`. The code below will fill the array `Numbers` with random numbers between 0 and 499. It creates a local object called 'Rand' and then uses the `Next` function to generate random numbers within a FOR loop. The two parameters passed to the `Next` function indicate the maximum range for the random numbers. The second number is exclusive and not included in the range.

```
Sub Main()
    Dim Rand As New Random
    Dim Numbers(9) As Integer

    For Counter = 0 To 9
        Numbers(Counter) = Rand.Next(0, 500)
    Next
End Sub
```

a Produce an array `Numbers` capable of holding 200 integer values.

b Use the random number generator to fill the array with a series of random positive integer values. All the random values should be less than 5000.

c Write a bubble sort algorithm to sort the values in the array into descending order and then output the values of the sorted array.

Chapter 11
Manipulating strings

IN THIS CHAPTER YOU WILL:

- understand how to create meaningful output messages
- understand how to convert text between upper and lower case
- understand how to use the index locations within a string to identify individual characters within that string
- understand how to iterate through all the characters in string.

Introduction

We use written language to pass on messages, express ideas or tell a story. No one can be entirely sure how many words exist in the English language but estimates in excess of 200 000 are not unusual. Certainly, words sit at the very heart of human communication.

It is therefore not surprising that many technology systems involve natural language, through manipulating and using words. Examples include the use of spell checking in documents, predictive text on mobile applications and the use of natural language on internet search engines. Visual Basic, in common with many programming languages, stores words and text using the string data type. Visual Basic also provides a series of features that allow the programmer to manipulate and use string variables to validate textual inputs or communicate effectively to the user.

11.1 Concatenation

It is usual to expect to be given some context to a value. When talking about distances, it is normal to state the units of measurement, such as 20 centimetres, 20 kilometres or 20 light years. Without the unit of measurement, the value of 20 is meaningless.

The term **concatenation** means 'to link together'. It is used to describe the way code can be used to create a message that consists of the linking of text and the values in variables.

In Visual Basic, the ampersand (the '&' symbol) is used to indicate concatenation. While the default pseudocode symbol for concatenation is the comma symbol.

> **KEY WORD**
>
> **concatenation:** to link together text and the values in variables to create a new string value.

DEMO TASK 11.1

A system takes in the title, first name and last name of a customer. As part of the process, the system needs to output the full name of the customer. For example, input values of:

Title	Mr
First Name	Richard
Last Name	Morgan

need to be output as a single string: Mr Richard Morgan.

> **CONTINUED**

Solution

The pseudocode code to achieve this is:

```
DECLARE Title : STRING ← ""
DECLARE FName : STRING ← ""
DECLARE LName : STRING ← ""
DECLARE FullName : STRING ← ""
OUTPUT "Enter your title"
INPUT Title
OUTPUT "Enter your first name"
INPUT FName
OUTPUT "Enter your last name"
INPUT LName
FullName ← Title . " " ,
FName , " " , LName
OUTPUT FullName
```

The Visual Basic code to achieve this is:

```
Dim Title As String = ""
Dim FName As String = ""
Dim LName As String = ""
Dim FullName As String = ""
Console.WriteLine("Enter your title")
Title = Console.ReadLine()
Console.WriteLine("Enter your first name")
FName = Console.ReadLine()
Console.WriteLine("Enter your last name")
LName = Console.ReadLine()

FullName = Title & " " & FName & " " & LName
Console.WriteLine(FullName)
```

Concatenation is also commonly used to give context to output values. In this example, the output value of `Area` has been calculated from the `Length` and `Height` values input by the user.

```
Area = Length * Height
Console.WriteLine("The area is " & Area & " square metres")
```

> **TIP**
>
> Because it is usual to see space between words, it is good practice to insert a space. This can be seen in the Area example as a space is included at the end of first text statement and the beginning of the last text statement.

> **PRACTICE TASKS 11.1–11.2**
>
> **11.1a** Design the pseudocode for an algorithm for a system that takes in as inputs the radius of a circle and the unit of measurement. The system will output the area of the circle using a specific message.
>
> For example, if the values input are `radius = 3.6` and `units = metres`, the message would read:
>
> 'The area of a circle with a radius of 3.6 metres is 40.72 square metres.'
>
> **b** Implement and test your algorithm using either a console application or Windows Forms application.
>
> **11.2** Work through some of your previous coded solutions to tasks. Improve the output message by using concatenation.

11.2 Splitting strings

Another common task required when working with text is to be able to split a single textual value into individual sections or even individual characters. The following sections show how either individual characters or groups of characters can be identified from an original string value. Some examples of how these approaches might be used in a program are shown in Table 11.1:

Explanation	Original Text	Required Element
Splitting email addresses to recognise the individual section or recipient.	info@cambridgeinternational.org	info
Separating the last four characters of a date to get the year.	22/01/2020	2020
Taking the first letter of a first name.	Richard Mr Richard Morgan	R Mr R Morgan

Table 11.1: Examples of splitting strings

11.3 String index

Each individual character in a string has a unique **string index** (see Table 11.2). All string values are numerically indexed, with the first index value being zero. Like arrays, strings use a **zero-based index**.

Index	0	1	2	3	4	5	6	7
String Value	C	o	m	p	u	t	e	r

Table 11.2: Indexing of characters in a string

Visual Basic has a range of inbuilt functions that use the index property of string variables and textual values to help the programmer identify individual characters.

Chars( ) function

The **Chars( )** function will return the character at the index indicated in a given string value. Table 11.3 shows some examples of the use of the Chars( ) function.

Visual Basic Code	Output

```Dim Word As String = "Computer"```
```Dim Letter As Char = ""```
```Letter = Word.Chars(4)```
```Console.WriteLine(Letter)```
	u
	This is at index 4 in the value in the string variable.

```Console.WriteLine("Computer Science".Chars(9))```	S This is at index 9 in the text value.

(continued)

> **KEY WORDS**
>
> **string index:** an index location given to each character within a string. The index is zero based from the first character.
>
> **zero-based index:** the index is numbered from zero. The first character in the string will be at index zero. The final index location will always be one less than the number of characters in the string.
>
> **Chars( ):** the function that will return the character at a given index location.

Visual Basic Code	Output

`Dim Word As String = "Hello"`
`Dim Letter As Char = ""`
`Letter = Word.Chars(5)`
`Console.WriteLine(Letter)`
	Will cause an 'out of index range error'. Although the string variable holds a five-letter word, the last index is 4. Remember zero-based index.

Table 11.3: Examples of the use of the Chars() function

IndexOf( ) function

The **IndexOf( )** function will return the index location of the first occurrence of a given character in a string value. If the given character does not appear in the string value, the function will return −1. Table 11.4 shows some examples of the `IndexOf( )` function.

> **KEY WORD**
>
> **IndexOf( ):** the function that will return the index location of a given character.

Visual Basic Code	Output

`Dim Word As String = "Alpha"`
`Dim Index As Integer = 0`
`Index = Word.IndexOf("a")`
`Console.WriteLine(Index)`
	4
	
	The character 'a' is at index 4 in the value in the string variable. `IndexOf()` is case sensitive. It will ignore the character 'A'.

`Console.WriteLine("$20.00".IndexOf("$"))`	$ The character '$' is at index 0 in the text value. `IndexOf` also works with symbols and numerical digits.
`Console.WriteLine("Gamma".IndexOf("m"))`	2 The first occurrence of the character 'm' is at index 2 in the text value. Only first occurrence is returned.

`Dim Word As String = "Beta"`
`Dim Index As Integer = 0`
`Index = Word.IndexOf("F")`
`Console.WriteLine(Index)`
	−1
	
	The character 'F' does not appear in the value in the string variable. A return of −1 is the default return for no match with the indicated character.

Table 11.4: Examples of the use of the Chars( ) function

UCASE( )

The ability to convert between case can be useful in converting user input, which could be a mix of case, into either uppercase or lowercase. The process is often involved when checking user input and consequently has its own pseudocode representation.

The `UCASE( )` pseudocode function will convert any string value to uppercase characters. Any symbols, numerical digits or characters already in upper case will remain unaltered.

Table 11.5 shows some examples of the pseudocode `UCASE( )` function.

Pseudocode	Output
`Word ← UCASE( "Computer Science 10")` `OUTPUT Word`	`"COMPUTER SCIENCE 10"`

`Word ← "Coding is FUN"`
`Word ← UCASE(Word)`
`OUTPUT Word`
	`"CODING IS FUN"`

Table 11.5: Examples of pseudocode `UCASE( )` function

LCASE( )

The `LCASE( )` pseudocode function will convert any string value to lowercase. Like the `UCASE( )` function, any symbols, numerical digits or characters already in the correct case will remain unaltered.

Table 11.6 shows some examples of the pseudocode `LCASE( )` function.

Pseudocode	Output
`Word ← LCASE( "COMPUTER Science 10")` `OUTPUT Word`	`"computer science 10"`

`Word ← "CODING IS FUN"`
`Word ← LCASE(Word)`
`OUTPUT Word`
	`"coding is fun"`

Table 11.6: Examples of pseudocode `LCASE( )` function

Visual Basic ToUpper and ToLower function

Not surprisingly, Visual Basic includes specific functions to convert between case.

Table 11.7 shows some examples of the `ToUpper` and `ToLower` functions:

Visual Basic Code	Output
`Console.WriteLine("Alpha1 Beta$".ToLower)`	alpha1 beta$ All letters changed to lower case. Symbols and numerical digits unaltered.

`Dim Word As String = "ABCdef 123+"`
`Dim UpperWord As String`
`UpperWord = Word.ToUpper`
`Console.WriteLine(UpperWord)`
	ABCDEF 123+
	
	The variable `UpperWord` holds an upper-case version of the data in the variable `Word`.

Table 11.7: Examples of the use of the `ToUpper` and `ToLower` functions

> CAMBRIDGE IGCSE™ & O LEVEL COMPUTER SCIENCE: PROGRAMMING BOOK

> **TIP**
>
> The `ToUpper` function can be useful when matching user inputs with required values where case is not important. Converting both values to upper case will allow a direct comparison as it will remove the complication of the user inputting values in a different case than the one expected. For example, inputting 'computer science' when the system expected 'Computer Science'.

Len( ) function

The **Len( )** function returns the length of the string value passed as a parameter. The length is the number of characters in the string. As strings use a zero-based index, the value of the length of a string will be greater than the value of the last index position. For example, the length of the word 'Hello' will be 5 but the last index location is 4. The 5 characters in the word are held in index positions 0 to 4.

This code will return the value 16:

```
Dim Word As String = "Computer Science"
Dim WordLength As Integer
WordLength = Len(Word)
Console.WriteLine(WordLength)
```

Remember that the space is a character. Although there are only 15 letters, with the space there are 16 individual characters.

> **KEY WORD**
>
> **Len( ):** the length function that can be used to return the number of characters within a string.

Pseudocode LENGTH ( ) function

Finding the length of a string is a common task and consequently a specific pseudocode representation of the `Len( )` function exists.

The pseudocode `LENGTH( )` is identical in operation to the `Len( )` function and will return the number of characters in a string.

As the pseudocode `LENGTH( )` and the `Visual Basic Len( )` seem very similar you might wonder why we have both. Remember not all languages have the same syntax for functions and while the `Visual Basic` syntax is similar to that of pseudocode, that may not be the case for other languages. When writing pseudocode remember to use the `LENGTH( )` syntax.

> **PRACTICE TASK 11.3**
>
> a   Implement and test a program that takes as an input a single string consisting of two words. The system will output the first letter of the first and second word in the string. The letters will be in upper case and separated by a space.
>
>   For example, if the input = 'Computer science', the output would be 'C S'.
>
> b   Explain what would happen if the user put a double space between the words.

> **TIP**
>
> Use the `IndexOf()` to find the space.

> **CHALLENGE TASK 11.1**
>
> a Design the pseudocode for an algorithm for a system that generates a password for a customer. The system will take the following inputs:
>
> - The first name of the customer.
> - The surname of the customer.
> - The year of birth of the customer as a four-digit number.
>
> The system will generate a password made up from: the first letter of last name in lower case followed by year of birth followed by the last letter of the first name in upper case.
>
> | First name | Bhavna |
> |---|---|
> | Last name | Kumar |
> | Year of birth | 2008 |
> | Password generated | k2008A |
>
> b Implement and test your algorithm using a Visual Basic program.

11.4 Substring

So far in this chapter we have learnt how to identify individual characters in a string. The `Substring( )` function allows a programmer to select a group of characters in the original string value, this is called **substringing**. The original string remains unaltered.

The `Substring( )` function takes two parameters in the format shown in Table 11.8:

Function	`'string variable'.Substring( x , y )`
Explanation	x is the index where the substring will begin
	y is the number of characters to substring

Table 11.8: Format of parameters in `Substring()`

> **KEY WORD**
>
> **substringing:** a programming method used to identify part of a string value by reference to the index locations within that string.

Table 11.9 shows some examples of the use of the `Substring()` function:

Visual Basic Code	Output
`Console.WriteLine("Alpha Beta".Substring(6, 4))`	Beta The substring has started from index 6 and taken four characters.

`Dim Word As String = "Computer Science"`
`Dim PartWord As String = ""`
`PartWord = Word.Substring(0, 8)`
`Console.WriteLine(PartWord)`
	Computer
	The substring has started from index 0 and taken eight characters from the variable `Word`. This value is stored in the variable `PartWord`.

`Console.WriteLine("Hello".Substring(3, 3))`	Will cause an 'out of index range error' as the substring is attempting to use index 3 to 6. The word 'Hello' only has index 0 to 5.

Table 11.9: Examples of the `Substring()` function

Pseudocode `SUBSTRING()` function

The pseudocode version of the `SUBSTRING()` function has an identical name but does operate slightly differently from the Visual Basic function.

The pseudocode version includes all the elements required to substring within the brackets. It has three different parameters. It is also common for pseudocode to index a string from 1 rather than the Visual Basic approach of indexing from 0.

Table 11.10 gives an explanation of the pseudocode `SUBSTRING()` function.

Pseudocode	Explanation
`SUBSTRING ("Computer Science", 1, 8)` will return "Computer"	The string value to be used is passed as the first parameter.
	The second parameter indicates the start of the substring. Because pseudocode indexes from 1 this is the first character.
	The third parameter is the number of characters required.
`Word ← "Coding is Fun"` `SUBSTRING (Word, 11, 3)` Will return "Fun"	Again, the first parameter is the string value to be used. In this example a string variable is passed.
	The start of the substring is indicated by the second parameter. As pseudocode indexes from 1 and spaces count as characters, the 11th character is the 'F' of 'Fun'.

Table 11.10: The pseudocode `SUBSTRING()` function

DEMO TASK 11.2

A system is required that takes a company email address and outputs two separate values. One value is the recipient and the other value is the company element of the email address.

Input – single email	Sales@thebusiness.org	
Output – as two separate elements	Sale	thebusiness.org

Solution

The approach to this task involves three steps:

Step 1 – Identify the index location of the @ symbol. In this example, index 5.

Step 2 – Substring from the start of the email (index zero) for the number of characters in the string up to the @ symbol. In this example, that is five characters – the same value as the index location of the @ symbol. As strings are zero-indexed, the number of characters before the @ symbol will always be the same as the index value of the @ symbol.

Step 3 – Substring from the index after the @ symbol for the number of characters following the @ symbol. To do this, you will need to use the length function to identify how many characters are in the whole string. In this example, there are 21 characters in the full email and we need to substring the last 15 characters to identify the company element of the full email. To arrive at the number of characters following the @ symbol, we could add 1 to the index value of @ symbol and then subtract that from the total number of characters.

Number of charters needed = total number of characters – (index of @ symbol + 1)

11 Manipulating strings

CONTINUED

The following is a possible pseudocode approach to the task:

```
DECLARE FullEmail : STRING ← ""
DECLARE Recipient : STRING ← ""
DECLARE Company : STRING ← ""
DECLARE AtIndex : INTEGER ← 0
DECLARE FullEmailLength : INTEGER ← 0
DECLARE CompanyCharacters : INT ← 0

OUTPUT "Enter full email address"
INPUT FullEmail
AtIndex ← FullEmail.IndexOf("@")
Recipient ← SUBSTRING (FullEmail, 1, AtIndex)
// Remember pseudocode defaults to indexing from 1
FullEmailLength ← LENGTH (FullEmail)
CompanyCharacters ← FullEmailLength - (AtIndex + 1)
Company ← SUBSTRING (FullEmail, AtIndex + 1, CompanyCharacters)
OUTPUT "The recipient of the email is " + Recipient
OUTPUT "The company element of the email is " + Company
```

The console application Visual Basic code would follow the same logic:

```vb
Sub Main()
        Dim FullEmail As String = ""
        Dim Recipient As String = ""
        Dim Company As String = ""
        Dim AtIndex As Integer = 0
        Dim FullEmailLength As Integer = 0
        Dim CompanyCharacters As Integer = 0

        Console.WriteLine("Enter full email address")
        FullEmail = Console.ReadLine()
        AtIndex = FullEmail.IndexOf("@")
        Recipient = FullEmail.Substring(0, AtIndex)
        FullEmailLength = Len(FullEmail)
        CompanyCharacters = FullEmailLength - (AtIndex + 1)
        Company = FullEmail.Substring(AtIndex + 1, CompanyCharacters)

        Console.WriteLine("The recipient of the email is " & Recipient)
        Console.WriteLine("The company element of the email is " & Company)
        Console.ReadKey()
    End Sub
```

> **CAMBRIDGE IGCSE™ & O LEVEL COMPUTER SCIENCE: PROGRAMMING BOOK**

CONTINUED

> **TIP**
>
> Coding each step in a process can help when identifying errors. It would have been possible to have coded the email task using fewer lines of code and variables:
>
> ```
> Recipient = FullEmail.Substring(0, FullEmail.IndexOf("@"))
> Company = FullEmail.Substring(FullEmail.IndexOf("@") + 1, Len(FullEmail) -
> (FullEmail.IndexOf("@") + 1))
> ```
>
> While this does create fewer lines of code, it does increase the need to use BODMAS accurately and makes the task of debugging errors more complex.

PRACTICE TASK 11.4

a   Design an algorithm in pseudocode for a system that takes as an input a person's first and last name and outputs the first name and second name as separate values in a message. For example, an input of 'Aditya Khatri' would produce an output of 'Your first name is Aditya and your last name is Khatri'.

b   Implement and test your algorithm using either a console application or Windows Forms application.

CHALLENGE TASK 11.2

Extend the name system to include a date of birth. The system will take a person's first and second name and their date of birth in the format dd/mm/yyyy. The system will output the first name, second name and the day and month of the person's birth as separate values in a message. For example, an input of 'Aditya Khatri 22/01/2010' would produce an output of 'Your first name is Aditya and your last name is Khatri. You were born in January of 2010'.

11.5 Looping through all the characters in a string

Programmers can be required to write code that will identify all the individual characters in a string. For example, a system that encrypts messages would be required to encrypt each individual character in the message.

FOR loops are often combined with substring commands to be able to achieve this. The Visual Basic code will follow this format:

```vb
Dim Message As String = ""
Console.WriteLine("Enter message")
Message = Console.ReadLine()

For counter = 0 To Len(Message) - 1
    Dim EachLetter As Char
    EachLetter = Message.Substring(counter, 1)
```

```
        'Code applied to EachLetter'
Next
```

A FOR loop is used to iterate through each character in the string variable 'Message'.

The loop counter starts at 0 (the first index location). The `Len( )` function is used to return the number of characters in the string. The final index is one less than the number of characters. So the loop ends at `Len(Message) - 1`.

The loop counter is used as the index location parameter in the substring command. As the loop counter increments, this value will be passed to the substring command and cause the loop to increment through all the index locations in turn. The number of characters parameter in the substring command is fixed at 1 so that at each iteration a single character is taken.

PRACTICE TASKS 11.5–11.6

11.5a Design the pseudocode of an algorithm for a system that counts the number of spaces in a text message. The system will take the text message as an input and output the number of spaces in the message.

   b Implement and test your algorithm using either a console application or Windows Forms application.

11.6a Design the pseudocode of an algorithm for a system that will take a text message and output the reverse of the message. For example, an input of 'Computer Science' would generate an output of 'ecneicS retupmoC'

   b Implement and test your algorithm using either a console application or Windows Forms application

SUMMARY

Concatenation offers the programmer a method of joining together text and the values in variables to create a new string value. This can be used to generate user-friendly output messages.
Substringing is used to allow the programmer to identify part of a string value by reference to the index locations within that string. It can be used to identify whole words within a string.
All the individual characters in a string can be referenced by an index location. In Visual Basic the index is numbered from zero and starts with the first character in a string. Spaces are characters, so will also have an index location. When using pseudocode it is common to index from 1.
Visual Basic includes a number of inbuilt string manipulation functions: • `Chars( )` returns the character at a given index location • `IndexOf( )` returns the first index location of a given character • `ToUpper` and `ToLower` alters the case of the string • `Len( )` returns the number of characters in a string.

END-OF-CHAPTER TASKS

1 a Design the pseudocode for an algorithm for a system that generates an email address for a business. The business email address is business.org. The system will take as inputs:
- The first name of the user.
- The last name of the user.

The system will output the business email in the format 'firstname.lastname@business.org'.

b Implement and test your algorithm using either a console application or Windows Forms application.

2 a Design the pseudocode for an algorithm for a system that sums all the digits in a number. The number input can be any size and include decimal values.

Number Input	Value Output	Process
14673	21	1 + 4 + 6 + 7 + 3
998.237	38	9 + 9 + 8 + 2 + 3 + 7
26	8	2 + 6

b Implement and test your algorithm using either a console application or Windows Forms application.

3 a Design the pseudocode of an algorithm for a system that checks for double spaces in a text message. The system takes as an input the text message. It will output the number of occasions a double space had been identified.

b Implement and test your algorithm using either a console application or Windows Forms application.

c Extend your system to also output the corrected message with all double spaces removed.

4 a Design the pseudocode of an algorithm for a system that generates a password for new users.

The password is generated based on these rules:
- The first character is the last letter of the first name in upper case.
- The second character is the first letter of the first name in lower case.
- The third character is the first letter of the first middle name in lower case. If the user does not have a middle name the character x is inserted.
- The fourth character is the number of letters in the user's last name.
- The final characters are the first four characters in the user last name all in upper case. If the user does not have four characters in their last name, then the full last name is used.

Name input	Username generated
David John Smith	Dsj5SMIT
Michael Alan Peter Lee	Lla3LEE
Aadarsh Chandra	Hcx7CHAN

b Implement and test your algorithm using either a console application or Windows Forms application.

Chapter 12
Programming scenario task

IN THIS CHAPTER YOU WILL:

- learn about how to walk though a scenario task
- learn how to decompose the scenario into Inputs, Processes and Outputs
- practise choosing meaningful names for all your variables
- practise fully commenting your code
- design data processing algorithms that are efficient and produce accurate results
- write programs that produce outputs in a form appropriate to the task.

CAMBRIDGE IGCSE™ & O LEVEL COMPUTER SCIENCE: PROGRAMMING BOOK

The information in this section is based on the Cambridge IGCSE, IGCSE (9–1) and O Level Computer Science syllabuses (0478/0984/2210) for examination from 2023. You should always refer to the appropriate syllabus document for the year of your examination to confirm the details and for more information. The syllabus document is available on the Cambridge International website at www.cambridgeinternational.org.

Introduction

Designing a complete system normally requires the the combination of many programming elements to ensure the final system meets the user requirements. In large commercial systems it is normal to have a team of designers and programmers working together to code and test the system. The complete system will have been broken down into individual tasks using a top-down design process, with the individual tasks being developed by different programmers.

Although you will be working independently, a programming scenario task will involve similar challenges. It will require you to break down the task into individual elements and then select appropriate programming approaches to complete each of the elements. Combining those individual elements will then produce a complete solution to the task. If there are areas that you don't understand, it might help to re-visit the relevant section of the book.

12.1 Getting started on the question

To respond to a scenario question you need to write your program in either pseudocode or program code. Using either method is fine, choose the method you are most comfortable using.

In Chapter 6, Designing algorithms, we looked at the four main components of a computer model:

Input → Storage → Process → Output

The first steps will be to identify each of these elements for the scenario question. Details of these steps are shown in Table 12.1:

Input	Not all scenario questions will involve inputs. However, if they are included, they will be described in the text of the question. Any inputs should be validated as part of your program.
Storage	The program is likely to require you to use variables, constants or arrays. Some questions may give you variable or array names that you must use. It is a good idea to highlight these in the question. When a team of programmers work together it is crucial that they use the same names for shared variables. Using different names in individual sections of code would mean that the code elements would not work together when combined. In a similar way it is crucial that you use any names given.

(continued)

> **KEY WORDS**
>
> **input:** data that is required by a program to complete the required task. The data can be obtained from the user, a file or database, a device or another program.
>
> **storage:** the use of data structures to temporarily or permanently store data. Common data structures are variables, constants and arrays.
>
> **process:** a part of a program where the input data is used to complete the required task.
>
> **output:** textual, visual or audio data that is produced by a program. The output can be sent to an output device, stored in a file or database or displayed to a user.

Process	Some processes will be fully described. For example, any conditions that you should follow will be accurately described. Be careful to make sure you follow the conditions exactly. For example, if the question indicates a condition of 'greater than or equal to 50', you will need to code `>=50`. Some processes may require you to work out what to do. In common processes such as calculating an average or outputting the highest value from an array you will need to select an appropriate approach to complete the task. It is best practice to become familiar with approaches to common tasks.
Output	The outputs will be fully described in the questions. These may also help you to decide what processes are required as you may have to code a process to produce the output required. Always use meaningful messages with your outputs. Use concatenation (see Chapter 11, Manipulating strings) to produce a message that includes the output value.

Table 12.1: The four main components of a computer model

12.2 Planning the program

It is not uncommon to start writing a program and then realise you have missed out some code.

It is best practice to plan the logic of the program before you start coding. This will help to ensure an appropriate flow to the program and reduce the change of missing out important elements of code. Remember that when writing pseudocode or programming code by hand you do not have an IDE to identify syntax errors. It is always best to spend a few minutes planning.

The following task is a worked example showing the planning and programming process that could be used.

DEMO TASK 12.1

A 1D array `EmployeeID[]` contains the unique identity numbers (IDs) of 50 employees in an organisation. Employee IDs are six characters long. A 2D array `EmpData[]` contains the year the employee joined the organisation and the employees current annual salary. The position of each employee's data in the arrays is the same. For example, the employee at position 10 in the array `EmployeeID[]` will have their data recorded at position 10 in the array `EmpData[]`.

The arrays have already been set up and all data stored.

Write a program that meets the following requirements:

- *allows a user to input an employee ID*
- *validates the input*
- *verifies the employee is in the array*

> **CONTINUED**
>
> - outputs for the employee with the input ID:
>   - the year they joined the organisation
>   - current annual salary.
> - outputs an error message if EmployeeID[] does not contain the ID input.
>
> You must use pseudocode or programming code and add comments to explain how your code works.
>
> **Solution**
>
> As this is an extended task, the rest of this chapter describes how to go about completing a full solution to this programming scenario task.

12.3 Planning your solution

Approach the task using computational thinking. Consider the Input, Storage, Process and Output requirements of the task (see Table 12.2).

Input	The user is required to input an Employee ID.
	This will need to be completed at the start of the process and should include some validation on the input.
	Clues to the expected validation are included in the words *'Employee IDs are six characters long'*. Should also include a presence check and length check on the input.
Storage	The data is already contained in the arrays EmployeeID[] and EmpData[]. It is crucial that these identifiers are used in the answer.
	The minimum size of these arrays can also be determined from the words in the question.
	The 1D array EmployeeID[] will be [0:49] because the questions states there are 50 employees. Data type of string.
	The 2D array EmpData[] will be [0:49, 0:1] or [0:1, 0:49] because it contains two data items (year joined and salary) for each employee. Data type of either integer or real.
	A variable is needed to hold the ID input by the user. There is no information in the question to indicate if the ID contains letters so suggest a STRING datatype.
Process	**Step 1:** Obtain and validate the user input.
	Step 2: Loop through EmployeeID[] array to search for the input ID.
	• You could use a count-controlled FOR loop to search all index locations in the array.
	• You could use a condition-controlled WHILE or UNTIL loop to search for the ID, stopping the search when the ID is located.
	UNTIL seems most appropriate option as loop must run at least once.
	Step 3: Within the loop, output the required data if ID matches.
	Step 4: Output 'Not found' message if ID is not in the array.

(continued)

Output	The output required is held in the `EmpData[]` array. You need a final output if no match is found. Must include meaningful messages with the outputs.

Table 12.2: The four main requirements

It is not expected that you would write down the considerations as they are shown here, but if it helps, you could make notes. Ensure these notes are clearly identified as such so they are not mistaken for part of your answer.

12.4 Writing your program

Start by declaring the arrays, variables or constants needed for the solution. Do not forget to check the spelling of any identifiers provided in the question and to comment your code. It is a good idea to leave some space before the input code just in case you have forgot to declare a variable you need later.

Commenting your program

It is important that you comment your code to explain the process and logic of your algorithm. Commenting is always best practice but in a scenario type task where there could be many ways to approach the task comments will help other people who might read your code to understand what is happening.

```
// declare variables needed
// arrays EmployeeID and EmpData have already been set up
   and data stored
DECLARE SearchID : STRING ← ""
```

Now to write the input code. Remember to include appropriate validation on all inputs.

```
// input and validation of ID by user
OUTPUT "Enter the ID to search"
INPUT SearchID
WHILE SearchID = "" OR LENGTH (SearchID) <> 6 DO
    OUTPUT "Please enter correct Employee ID"
    INPUT SearchID
ENDWHILE
```

Next, produce the search and output code using the REPEAT...UNTIL loop. This will require a loop counter to stop the loop when it reaches the end of the array, and a Boolean variable to indicate if a match has been found. These can either be declared just before the loop or with the other declarations at the start of the code.

```
DECLARE Counter : INTEGER ← 0
DECLARE Found : BOOLEAN ← False
// Loop to search EmployeeID[] for matching ID. IF match is found,
   related data in EmpData[] is output
REPEAT
    IF SearchID = EmployeeID [Counter]
      THEN
```

```
            OUTPUT "Year joined: ", EmpData [Counter, 0]
            OUTPUT "Current Salary", EmpData [Counter, 0]
            Found ← TRUE
        ENDIF
        Counter ← Counter + 1
    UNTIL Found = TRUE OR Counter = 50
```

Finally, generate the output that will be displayed if no match is found for the input ID.

```
// error message if no match found
IF Found = FALSE
  THEN
      OUTPUT "The ID you input is not in the records."
ENDIF
```

The complete pseudocode is shown here:

```
// declare global arrays and variables needed
// arrays EmployeeID and EmpData have already been set up and data stored
DECLARE SearchID : STRING ← ""
// input and validation of ID by user
OUTPUT "Enter the ID to search"
INPUT SearchID
WHILE SearchID = "" OR LENGTH (SearchID) <> 6 DO
    OUTPUT "Please enter correct Employee ID"
    INPUT SearchID
ENDWHILE
DECLARE Counter : INTEGER ← 0
DECLARE Found : BOOLEAN ← False
// Loop to search EmployeeID[] for matching ID, IF match found related data in
EmpData[] is output
REPEAT
    IF SearchID = EmployeeID [Counter]
      THEN
         OUTPUT "Year joined: " , EmpData [Counter, 0]
         OUTPUT "Current Salary" , EmpData [Counter, 0]
         Found ← TRUE
    ENDIF
    Counter ← Counter + 1
UNTIL Found = TRUE OR Counter = 50
// error message if no match found
    IF Found = FALSE
      THEN
         OUTPUT "The ID " , SearchID , " is not in the records"
    ENDIF
```

The same algorithm written as a console application would look like this:

```
Sub Main()
    'declare global arrays and variables needed.
    'Arrays EmployeeID and EmpData declared elsewhere in the code

    Dim SearchID As String
    'input and validation of ID by user
    Console.WriteLine("Enter the ID to search")
```

```vb
    SearchID = Console.ReadLine()
    Do While SearchID = "" Or SearchID.Length <> 6
        Console.WriteLine("Enter correct ID to search")
        SearchID = Console.ReadLine()
    Loop


    Dim Counter As Integer = 0
    Dim Found As Boolean = False
    'loop to search array EmployeeID for matching ID.
    'If match Then output related data in array EmpData
    Do
        If SearchID = EmployeeID(Counter) Then
            Console.WriteLine("Year joined: " &
EmpData(Counter, 0))
            Console.WriteLine("Current salary: " &
EmpData(Counter, 1))
            Found = True
        End If
        Counter = Counter + 1
    Loop Until Found = True Or Counter = 50
    'error message if no match for ID
    If Found = False Then
        Console.WriteLine("The ID " & SearchID & " is not
in the records")
    End If
    Console.ReadKey()
End Sub
```

12.5 Checking your program

You will not be able to code and test the application to identify errors. But you can check the program by reading through the code. A second read of the code can help identify mistakes you may have made when writing the code. Some common areas for mistakes are:

- Forgetting to declare variables.

- Misspelling the names of arrays or variables that have been provided in the question.

- Starting or ending loops with the wrong value. For example, in the pseudocode solution, we indexed the arrays from 0. Starting the loop from 1 would be an error and would mean the first index location was never searched. Another possible error could be setting the exit condition of the UNTIL loop to be when the counter was 49 as this would mean the last index location was never searched. As the counter is incremented before the loop, the check would need to be one greater than the last index position. This was achieved in this example by setting the loop limit to 50. Another approach would have been to set the loop to >49.

- Using incorrect conditions for your loops. For example, if we had incorrectly written the AND connector in the REPEAT...UNTIL loop, the code would only run once as the variable `Counter` would only hold the value of 1 when the condition was checked.

- Using incorrect logical operators. For example, if we had incorrectly written the validation condition as `SearchID.Length > 6`, the code would not work as required.
- Forgetting to comment your code and give meaningful messages with outputs.

12.6 Final thoughts

Once you have finished what you can, it is useful to go through a mental checklist to try to ensure you get as many of the marks available as possible. Table 12.3 shows a suggested checklist.

Variables:	<ul><li>Have I used the variable names provided?</li><li>Are my own variable names meaningful?</li><li>Are my variables all of the correct data type?</li></ul>
Inputs	<ul><li>Have I included effective validation for all inputs?</li></ul>
Processes:	<ul><li>Are my algorithmic solutions efficient?</li><li>Have I used appropriate loops?</li><li>Are the data structures I have used in my algorithms the best ones to use?</li><li>Do my data structures store all the data they should?</li></ul>
Outputs:	<ul><li>Does my program produce all the outputs required?</li><li>Are the outputs in the form asked for?</li></ul>
Other:	<ul><li>Is my solution properly commented?</li></ul>

Table 12.3: Mental checklist

SUMMARY

Identify the inputs, processes and outputs of the scenario.
Use appropriate messages when inputting and outputting data.
Create a structured plan using commenting.
Use meaningful identifiers for variables, constants and subroutines.
Select appropriate data types if not provided.
Choose efficient algorithms for processing data.
Ensure outputs are in the form requested.
Add further comments to explain your code, if necessary.

END-OF-CHAPTER TASKS

The end-of-chapter tasks provided here require similar approaches to the demonstration task you have just studied. By completing them, you will have used a range of programming techniques as well as developed your own approach to answering scenario questions. Once you are competent at completing scenario task you should aim to complete these types of task in 30 minutes.

1 A 1D array `Schools[]` contains the names of 30 schools taking part in a weather-monitoring project. The project lasts for 7 days. Each school records both the minimum and the maximum temperature, in Celsius, at their location for each of the 7 days. The 2D array `MinTemp[]` contains the minimum temperature recorded at each school for each of the 7 days. The 2D array `MaxTemp[ ]` contains the maximum temperature recorded at each school for each of the 7 days. The position of each school's data in the three arrays is the same.

   The arrays and variables have already been set up and the data stored.

   Write a program that meets the following requirements:
   - Calculates the daily temperature range (maximum temperature minus minimum temperature) for each of the schools.
   - Calculates the average daily temperature range over the 7-day period for each school.
   - Calculates the average maximum temperature over the 7-day period for each school.
   - Outputs, for each school:
     - name
     - average temperature range
     - average maximum temperature.
   - Calculates, stores and outputs the highest temperature and the lowest temperature recorded at any of the schools.

   You must use pseudocode or programming code and add comments to explain how your code works.

   You do not need to initialise the data in the arrays.

2 A bank is developing a system for handling interactions with customers. You are requested to provide the procedure `IDCheck( )` that is to be used to validate a customer identity number.

   A valid customer identity number consists of four lower case letters (a–z) followed by two numeric digits (0–9).

   Write the procedure `IDCheck( )` that meets the following requirements:
   - Requests the input of the customer identity number.
   - Checks the customer identity number input is in a valid format.
   - Outputs an error message and prompt for re-entry of the customer identity number when an invalid entry is made.
   - Allows a maximum of three attempts at making a valid input before outputting the message 'No further attempts'.
   - When a valid customer identity number is entered, the program will call the procedure `PasswordCheck(ByVal CustID)` and pass the customer identity number as a parameter.

   You must use pseudocode or programming code and add comments to explain how your code works.

CONTINUED

3 A 1D array `Members[]` contains the name of 50 members of a running club. All members of the club have taken part in a competition. The distance each member could run in 30 minutes was recorded. A 2D array `Distance[]` contains the distance, in metres, for each run for each member. It was possible for members to have three attempts at the run but not all members completed three runs. The value −1 has been recorded where the member did not complete a run. The position of each member's data is the arrays is the same. For example, the member at position 20 in the array `Members[]` will have their runs recorded at position 20 in the array `Distance[]`.

The arrays have already been set up and the data stored.

Members are allocated categories based on the distance achieved in their best run:

Category	Distance achieved in best run
Elite	Greater than or equal to 7 kilometres.
Championship	Greater than or equal to 5 kilometres and less than 7 kilometres.
Club	Less than 5 kilometres.

Write a program that meets the following requirements:
- Calculates the average distance run for each member. Where a member has completed fewer than three runs, the average will be based only on the number of runs, they completed.
- Calculates the best distance for each member.
- Outputs, for each member:
  - name
  - number of runs completed
  - average distance covered
  - best distance
  - category awarded.
- Calculates, stores and outputs the total distance ran by all competitors.

You must use pseudocode or programming code and add comments to explain how your code works.

You do not need to initialise the data in the arrays.

4 A program is needed to help young children practise basic mathematical addition skills.

The program will challenge the child to input correct answers to 10 randomly generated addition questions. The program will display each question to the child and prompt for the child to enter the correct answer to the calculation.

At the start of the program, the child will be prompted to input the level of questions they wish to be asked.

Level of question	Type of random question generated
Easy	Each question will consist of two random whole numbers between 1 and 20.
Medium	Each question will consist of two random whole numbers between 1 and 50.
Hard	Each question will consist of two random whole numbers between 1 and 100.

Write a program that meets the following requirements:
- Prompts the child to select a level for the questions.
- Outputs a randomly generated question.
- Prompts the child to input an answer.

CONTINUED

- Repeats the process 10 times.
- At the end of the 10 questions, outputs:
  - the number of correct answers
  - all the questions the child answered incorrectly with the correct solution.

You must use pseudocode or programming code and add comments to explain how your code works.

5 A traffic survey has recorded the daily number of cars, heavy vehicles and motorcycles using a road over a 5-day period.

An example of the data recorded is shown below:

	Car	Lorry	Motorcycle
Day 1	11400	6012	2890
Day 2	13280	1407	4278
Day 3	23375	5612	2965
Day 4	12786	3256	3872
Day 5	26356	7023	3260

A 2D array `Vehicles[]` is to be used to hold the five daily values recorded for each type of vehicle.

Write a program that meets the following requirements:

- Declares the array `Vehicles[]`.
- Prompts for the daily values recorded for each type of vehicle to be entered into the array.

Once the input has been completed the program will:

- Calculate and output the total number of vehicles using the road over the 5-day period.
- Calculate and output the number of days that the daily number of vehicles exceeded 20 000.
- Calculate and output the average daily number of cars.
- Calculate and output the average daily number of lorries.
- Calculate an output the average daily number of motorcycles.

You must use pseudocode or programming code and add comments to explain how your code works.

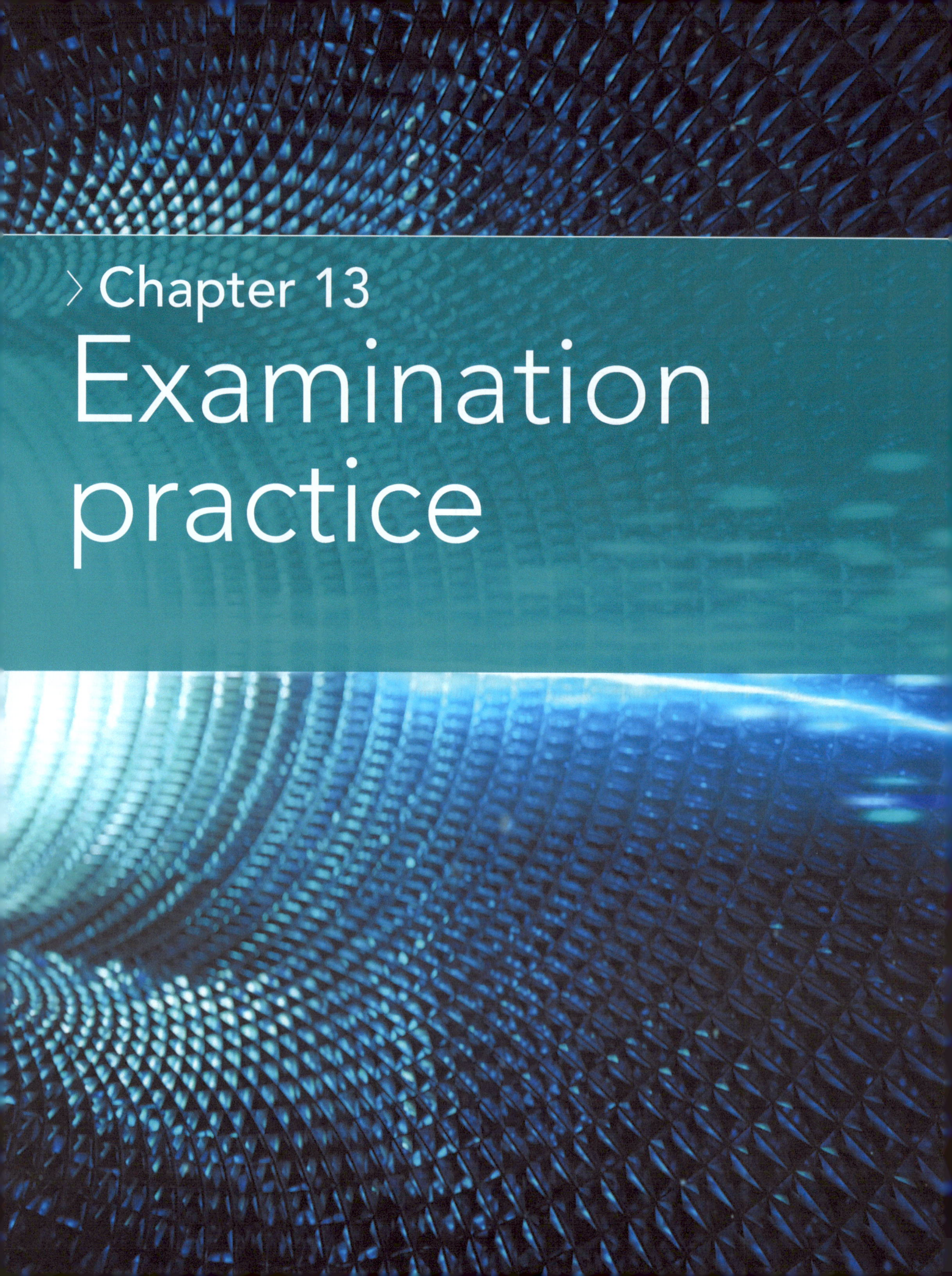

Chapter 13
Examination practice

13 Examination practice

Exam-style questions and mark scheme guidance have been written by the authors. In examinations, the way marks are awarded may be different. References to assessment and/ or assessment preparation are the publisher's interpretation of the syllabus requirements and may not fully reflect the approach of Cambridge Assessment International Education.

Introduction

This chapter includes a series of exam-style questions. This chapter aims to bring together all the skills you have developed and all the knowledge you have gained throughout this book. The questions will test your understanding of key programming concepts you have learned and offer you an opportunity to identify any gaps in your understanding. If you are unsure about how to answer a particular question, it may help to re-visit the relevant section of the book. There are full, suggested solutions in the solutions chapter in the digital part of this resource. Remember that there is often more than one solution. It is important that you comment your code to explain the logic of your algorithms. Good luck!

EXAM-STYLE QUESTIONS

1   A school is holding a vote to decide on the name for a new science block. The two options are Faraday and Curie. Each of the 601 students will vote. An algorithm is to be created for an electronic voting system that will record the vote of each student. You can assume that all the students will make a valid vote. The system will output the most popular name and the percentage of students that voted for that name.

   Write an algorithm for the voting system, using pseudocode or a flowchart. [6]

2   The pseudocode inputs the scores achieved by students in a test. The maximum score in the test was 200, A value of −1 stops the input. The algorithm outputs the highest, lowest and average score.

```
DECLARE Marks : INTEGER . 0
DECLARE High : INTEGER . 0
DECLARE Low : INTEGER . 0
DECLARE Total : INTEGER . 0
DECLARE Count : INTEGER . 0
DECLARE Average : REAL . 0
OUTPUT "Enter Student Marks"
INPUT Marks
WHILE Marks <> -1 DO
    Total ← Total + Marks
    Count ← Count + 1
    IF Marks <= HIGH
      THEN
         High ← Marks
    ELSE IF Marks < Low
      THEN
        Marks ← Low
    ENDIF
       OUTPUT "Enter Student Marks"
       INPUT Marks
ENDWHILE
Average ← Count / Total
OUTPUT High , " " , Low , " " , Average
```

> **CONTINUED**

There are four errors in the algorithm.

Locate those errors and show how the error could be corrected. [8]

3 A system accepts 13-digit ISBN numbers. The ISBN number includes a check digit.

   a Explain how a check digit is used to validate the input [4]

   b The pseudocode has been written to complete other types of validation. For each example state what type of validation is being completed.

Pseudocode	Type of validation

`INPUT ISBN`
`IF ISBN = ""`
`   THEN`
`      OUTPUT "Error"`
`ENDIF`	

`IF LENGTH (ISBN ) < 13`
`   THEN`
`      OUTPUT "Error"`
`ENDIF`	

[2]

   c This pseudocode has been written to validate user input for a system that accepts only integer values.

```
DECLARE UserValue : INTEGER ← 0
INPUT UserValue
WHILE UserValue <=10 OR UserInput > 200 DO
    OUTPUT "Out of range, re-enter value"
    INPUT UserValue
ENDWHILE
```

The code is to be tested using normal, abnormal and extreme data.

Give one example of each type of test data (Normal, Abnormal and Extreme) that could be used to test the validation algorithm. [3]

4 The algorithm inputs a series of integer values. A negative value stops the input.

```
DECLARE Number : INTEGER ← 0
DECLARE Count : INTEGER ← 0
DECLARE Total : INTEGER ← 0
DECLARE Large : INTEGER ← 0
REPEAT
    INPUT Number
    Count ← Count + 1
    IF Number > 10
       THEN
          Total ← Total + Number
          IF Total MOD 20 = 0
             THEN
                Large = Total DIV 20
          ENDIF
       ENDIF
UNTIL Number < 0
Count ← Count - 1
OUTPUT Large * Count
```

CONTINUED

Create and complete the trace table for the following sequence of inputs:
15, 10, 14, 0, 12, 19, 11, −6.

Number	Count	Total	Large	Output

[5]

5   An array `Netname` holds the network username of 600 employees. The array has not been sorted.

An algorithm is required that will search the array for a specific username. If the username is in the array the algorithm will output the index location of the array that holds the specific username. If the username is not in the array the algorithm will output the message 'No Record".

   a   Write the search algorithm, using pseudocode or a flowchart. [5]

   The array `Netname` has been sorted in alphabetical ascending order.

   b   Explain how your algorithm could be made more efficient now the array `Netname` is sorted. [3]

6   A two-dimensional array is used in an electronic board game. The pseudocode to declare the array is:

```
DECLARE Board : ARRAY [0:8, 0:8] OF STRING
```

When each game starts each index location in the array should be assigned the value "Free".

Write a pseudocode an algorithm to assign the value "Free" to all index locations in the array `Board`. [3]

7   A business requires that product numbers meet these requirements:

   • The product number must begin with the letters "PROD".

   • The product number must be 10 characters long.

The product numbers PROD1556DT and PRODa45629 are valid product numbers.

A system is required that will check that product numbers meet the requirements. The system will output the message "Accepted" or "Rejected" depending on whether the product number input meets the rules.

Write the algorithm the will that will check that product numbers meet the requirements. Use pseudocode or a flowchart. [6]

8   Rewrite the following pseudocode using a CASE statement.

```
DECLARE Score : INTEGER ← 0
INPUT Score
IF Score > 100
  THEN
    OUTPUT "Excelling"
  ELSE
    IF Score > 80
      THEN
        OUTPUT "Good"
      ELSE
        IF Score > 60
          THEN
            OUTPUT " Acceptable"
          ELSE
            OUTPUT "Below expectations"
        ENDIF
    ENDIF
ENDIF
```

[4]

> **CONTINUED**
>
> **9** A function `convert` is required which will take as a parameter the temperature in Celsius and returns the equivalent temperature in Fahrenheit.
>
> The formula to convert from Celsius (C) to Fahrenheit (F) is F = C * 1.8 + 32.
>
> Write the pseudocode for the function `convert`. [2]
>
> **10** Programmers make use of variables and constants.
>
> **a** Explain **one** difference between a variable and a constant. [2]
>
> Variables and constants are declared using data types.
>
> **b** Provide the most appropriate data types for each variable:
> - `Students` – holds the number of students in a school
> - `IDNumber` – holds ID numbers. An example is MR145T
> - `Weight` – holds the weight of parcels in kilograms
> - `Passed` – records whether a person has passed a test
> - `Name` – holds the name of students at a school
> - `Grad` – holds an examination grade A, B, C, D or E. [6]
>
> **11** A tennis club classifies its members as follows:
> - child if aged less than 11
> - junior if aged 11 to 17
> - adult if aged 18 to 59
> - senior if aged 60 or older.
>
> The following is a partially completed flowchart design for a program which would be used to store the classification of each member in a variable called classification. [8]

CONTINUED

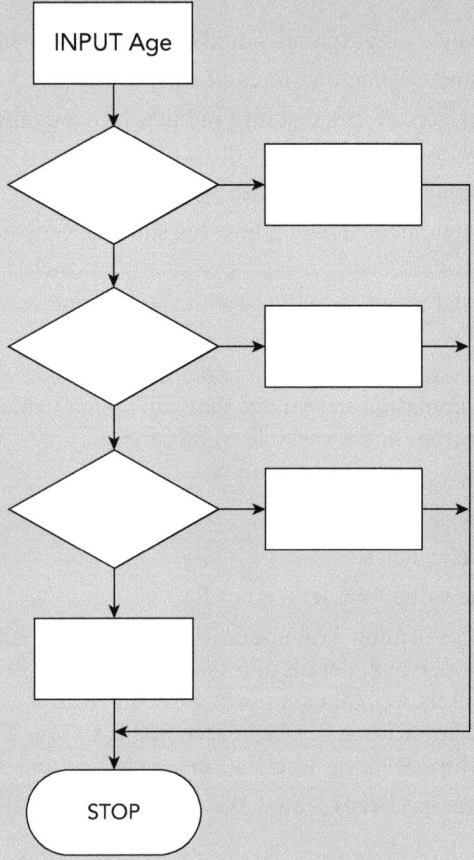

Complete the flowchart to show the design of an appropriate algorithm for this task.

12 a Define the term **variable**. [2]

b A school stores data in a computer system regarding students performance in tests. The following table shows some example data that would be stored by the system:

Student Name	Student identifier	Mark	Percentage	Comments	Re-test needed
A Patel	346	60	80.5	improved	N
W Chi	231	65	87	steady	N
E Brown	458	33	44	disappointing	Y
M Santana	023	42	56.5	better	N

Identify the data type you would use for each data element (heading of each column in table above). [6]

CONTINUED

13 a i State three reasons why a programmer would choose to use a subroutine. [3]

　　ii State two differences between a function and a procedure. [2]

　b A subroutine is to be used to convert a distance in miles into a distance in kilometres. The calculated value is stored in a variable with the identifier `KmDistance`.

　　i Choose a suitable name for the subroutine.

　　ii Write a pseudocode statement showing how the subroutine would be used if it had been written as a procedure.

　　iii Write a pseudocode statement showing how the subroutine would be used if it had been written as a function. [3]

14 A program contains a variable `oldString` which holds the value 'someWORD'. Using pseudocode or program code write string manipulation statements that will assign values to another variable called `newString` from the value current in the variable `oldString`.

　a `newString` is to hold the value "SOMEWORD"

　b `newString` is to hold the value "D"

　c `newString` is to hold the value "s"

　d `newString` is to hold the value "WORD" [4]

15 A club is developing a system for dealing with applications for membership. When people apply to join the club they are asked for personal details and the name of an existing club member who will recommend them. The club will then obtain supporting details from the existing club member. The decision as to whether the application has been accepted is stored. The applicant will be sent either a 'Welcome to membership' or 'Sorry application rejected' letter.

The following structure diagram has been created as a design for the system. You need to provide labels for each box in the diagram.

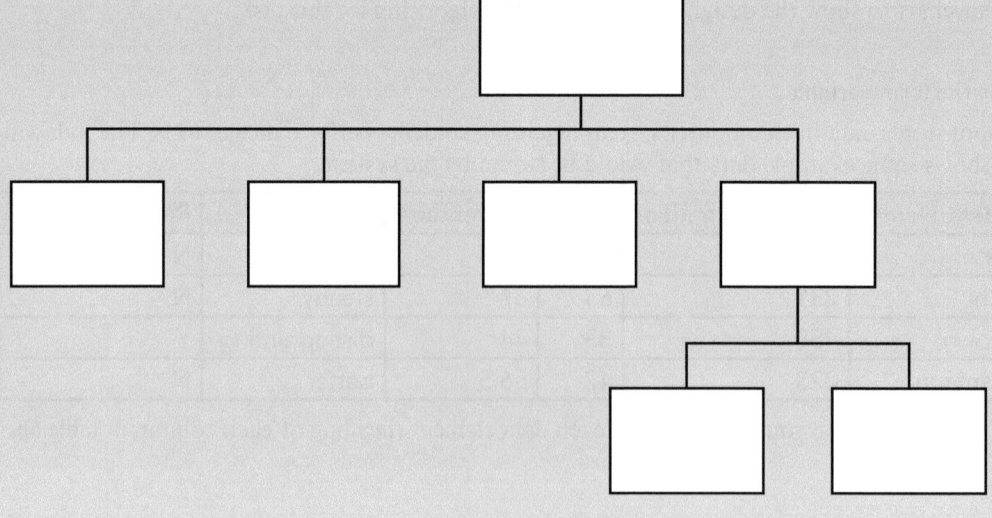

[5]

CONTINUED

16 Consider the following fragment of pseudocode:

```
1.  DECLARE Swaps : BOOLEAN ← True
2.  DECLARE CounterLimit : INTEGER ← 0
3.  DECLARE Size : INTEGER ← 0
4.  Size ← array.Length
5.  WHILE Swaps = TRUE DO
6.      Swaps ← FALSE
7.      FOR Counter ← 0 TO ((Size - 2) - CounterLimit))
8.          IF array[Counter] > array[Counter + 1]
9.              THEN
10.                 Temp : INTEGER ← 0
11.                 Temp ← array [Counter]
12.                 array [Counter] ← array [Counter + 1]
13.                 array [Counter + 1] ← Temp
14.                 Swaps ← TRUE
15.             ENDIF
16.     NEXT Counter
17.     CounterLimit ← CounterLimit + 1
18. ENDWHILE
```

    a Identify the purpose of this algorithm. [1]

    b Identify the line numbers where this code uses the following programming constructs:

- iteration
- selection. [3]

    c The code at line 5 reads `WHILE Swaps = TRUE DO`

       Explain the role of the WHILE loop in this algorithm. [3]

    d The code includes the variable `CounterLimit`.

       Explain how this variable is used to improve the efficiency of this algorithm. [2]

17 A student writes a program for a running club. The program requires the user to input the following information:

- First name and last name.
- The time it took to complete each of the last 5 marathons. Each time is rounded to the nearest minute.

The program will output the average time for the 5 marathon runs. The average is output in hours and minutes.

This is the pseudocode for the algorithm:

```
FUNCTION f (i : REAL, j : REAL, k : REAL, l : REAL, m : REAL)
RETURNS REAL
    t ← i + j + k + l + m
    RETURN ROUND(t/5)
ENDFUNCTION
INPUT name1
INPUT name2
FOR Counter ← 1 TO 5
```

CONTINUED

```
      CASE OF i:
         1: INPUT a
         2: INPUT b
         3: INPUT c
         4: INPUT d
         5: INPUT e
      ENDCASE
NEXT Counter
ave ← f(a,b,c,d,e)
OUTPUT name1, " ", name2, " average marathon time:"
OUTPUT ave DIV 60 , "hrs " , ave MOD 60 , "mins"
```

a Re-write the algorithm, keeping its logical flow, but make it easier to read and more manageable. [6]

b Another programmer tells the student that the algorithm is overly complex and could be produced without the use of the FUNCTION or the CASE statement. Rewrite the algorithm without using FUNCTION or CASE. [6]

18 A simple game app keeps track of the number of lives a player has in a variable called lives. Players can either gain or lose lives. Depending on the event that happens, one of the two procedures in the code snippet below are called.

```
DECLARE Lives : INTEGER ← 5
DECLARE GameOver : BOOLEAN ← False
PROCEDURE loseLife()
    MinLives ← 0
    IF lives > MinLives
      THEN
         lives ← lives - 1
    ENDIF
    IF Lives = 0
      THEN
         GameOver ← TRUE
    ENDIF
ENDPROCEDURE
PROCEDURE gainLife()
    MaxLives ← 5
IF Lives < MaxLives
  THEN
     Lives ← Lives + 1
ENDIF
ENDPROCEDURE
```

a Give the identifiers of all the local variables. [1]

b Give the identifiers of all the global variables. [1]

c Explain the difference between a local and global variable. [2]

CONTINUED

19 A program is required that creates usernames from users' first and last names. First the program must ask a user to input their first name and then their last name. It then creates a username from the first two letters of the first name and the last five letters of the last name. The username is entirely lowercase. The program will finally output the username

Assumptions:
- Ignore possible duplicate usernames.
- Users' first and last names only contain letters (no hyphens or apostrophes or accents).
- Users' first and last names may contain a mixture of uppercase and lowercase letters.
- Users' first and last names always contain more than 3 letters and 5 letter respectively.

Write an algorithm, using pseudocode or program code, to perform this part of the larger program. [6]

20 A simple computer game currently stores a map of rooms in a 10 × 10 grid in ten arrays like this:

```
DECLARE row0 : ARRAY [0 : 9] OF STRING
DECLARE row1 : ARRAY [0 : 9] OF STRING
DECLARE row2 : ARRAY [0 : 9] OF STRING
```

etc.

a Why would it be better to store the map data in a single two-dimensional (2D) array? [3]

b Provide the declaration pseudocode for a 2D array, `map`, that can be used to store all the information in the ten `1D` arrays currently used. [1]

c The file name of an image is stored in `map[2][2]`. Stored in `map[4,4]` is the file name of another image. During the game these two image file names need to be swapped around. Write an algorithm using pseudocode that would swap the two image references in the map array. [3]

Glossary

algorithm: a process or set of rules to be followed during the execution of a program.

alpha testing: early testing that takes place during the programming of a system.

AND Boolean operator: used to create a condition which is made up from more than one criteria. The condition will only be TRUE if **all** of the individual criteria are TRUE.

array: a variable that can hold a set of data items of the same data type under a single identifier.

array index: a series of sequential numbers that reference the data items in an array. In Visual Basic, the first index value is zero. This is known as a zero-based index.

beta testing: formal testing once the system has been completed.

breakpoint: points that are inserted into the code by the programmer. They will cause the system to run one line of code at a time once the breakpoint is reached. Helps the programmer to track the value of variables as the program executes.

bubble sort: an algorithm that is used to sort the values in an array into ascending or descending order.

calling: to activate a subroutine. To do this, you specify the subroutine's name and, optionally, parameters.

CASE statement: a simple method of providing multiple paths through the code based on a single variable or user input.

Chars( ): the function that will return the character at a given index location.

classes: templates that hold prewritten code which supports the functionality of objects. All of the GUI elements in Visual Basic are objects of a class. The GUI objects can be created and manipulated without the need to write code as the code required is already contained in the class.

command-line interface: an interface that uses text on a single screen to input data into a system and output information from that system.

commenting: a description of the algorithm written within the code. The comments are intended to help explain how the code works. Comments are ignored by the computer when the code is executed.

computational thinking: the thought processes involved in expressing solutions as computational steps or algorithms that can be carried out by a computer. These include abstraction, problem analysis, step-wise refinement, decomposition and algorithm design.

concatenation: to link together text and the values in variables to create a new string value.

condition: the criteria that are tested as part of the execution of the code. The condition will result in either a True or False answer when tested.

condition-controlled loops: types of iteration where the repetition of the loop is determined by conditions. The amount of times the loop will be executed is unknown.

console application: a command-line interface provided by Visual Studio 2019 Community used to create programs using the Visual Basic programming language. This interface most directly matches the syllabuses.

constant: a named memory location used to store a value; the value can be used but not changed during program execution.

construct: a method of controlling the order in which the statements in an algorithm are executed.

criteria: the specific rules that are used by the program to recognise if a condition is TRUE or FALSE. The singular form of criteria is criterion.

data integrity: the correctness of data during and after processing.

debugging: a general term for systematically searching for problems in programs that are not working properly and fixing them.

declaring: setting up a variable or constant.

decomposition: a computational thinking skill that involves thinking about large tasks and breaking them down into smaller tasks.

diagnostics: the systematic process of trying to diagnose what is wrong with a program.

ELSEIF statement: an alternative to nested IF statements. Each condition is tested in turn. When a condition is TRUE, the code related to that condition is executed and the statement is ended.

Glossary

execute: in Computer Science, the term 'execute' means the operation of a computer program. When a computer program is in operation it is being executed. The term 'run' is also used to describe the same process. The 'program is running' or the 'program is being executed' both mean the same thing.

file handling: programming statements that allow text files to be opened, read from, written to and closed.

flowchart: a graphical representation of the sequence and logic of a program.

FOR loop: a type of iteration that will repeat a section of code a known number of times. Also known as a count-controlled loop.

function: a subroutine that can receive multiple parameters and returns a single value. A function always returns a value through its identifier.

global variable: a variable that can be accessed from any routine within the program.

Graphical User Interface (GUI): an interface that includes graphical elements, such as windows, icons and buttons.

identifier: the name given to each variable or constant. The identifier is used to reference the memory area.

IF statement: a statement that allows a program to follow or ignore a sequence of code depending on a Boolean condition.

IndexOf( ): the function that will return the index location of a given character.

infinite loop: a loop that will iterate indefinitely. Usually caused by a logical error in the condition that controls the loop.

initialising: giving a variable a start (initial) value when it is first declared.

input: data that is required by a program to complete the required task. The data can be obtained from the user, a file or database, a device or another program.

Integrated Development Environment (IDE): software that helps programmers to design, create and test program code.

interface: the way in which a user inputs data into and receives information from a computer system.

iteration: code repeats a certain sequence a number of times depending on certain conditions.

Len( ): the length function that can be used to return the number of characters within a string.

linear search: a method of searching an array. The process checks every element in the array sequentially until the required element is found or the end of the array is reached.

local variable: a variable that can only be accessed in the code element in which it is declared.

logical errors: result in code that runs but produces unexpected results.

loop counter: a variable that is used within a FOR loop to keep a record of the amount of times the loop has been repeated. The loop counter normally increases by 1 each time the loop is executed.

metadata: data about data; information about the structure or format of the data stored.

nested IF statement: an IF statement with the ability for additional conditions to be checked once earlier conditions have determined a path.

NOT Boolean operator: used to create a condition which will be TRUE when the criteria are FALSE.

one-dimensional array: a linear array with a single index set. Contains one row of data with multiple elements; each element is identified by a unique index number.

OR Boolean operator: used to create a condition which is made up from more than one criteria. The condition will be TRUE if **any** of the individual criteria are TRUE.

output: textual, visual or audio data that is produced by a program. The output can be sent to an output device, stored in a file or database or displayed to a user.

overflow errors: occur when the data passed to a variable is too large to be held by the data type selected.

parameter: data or values that are passed to, or received from, a subroutine.

PascalCase: a way of creating a variable name from a combination of at least two words. Each new word starts with a capital letter.

passing by reference: the subroutine is passed a value by reference to a variable or array declared in the main code that holds the data to be used. Any changes made to that value is stored in the referenced variable in the main code.

passing by value: the subroutine holds a local copy of the data passed. Any changes made to the data are held in the local copy within the subroutine. If the data passed originated from a variable in the main code, the original value in that variable remains unaltered.

procedure: a subroutine that can receive and return multiple parameters. It may or may not return a value. If values are returned, they are returned via parameters.

process: a part of a program where the input data is used to complete the required task.

pseudocode: a way of unambiguously representing the sequence and logic of a program using both natural language and code-like statements.

REPEAT...UNTIL loop: a type of iteration that will repeat a sequence of code until a certain condition is met. The code within the loop will always be executed at least once. They are also known as post-condition loops.

reading data from a file: the process of reading the data held in the text file.

runtime errors: problems with the code that only become evident when the program is run, for example, attempting to divide by zero.

selection: code follows a different sequence based on what condition is chosen.

sequence: code is executed in the order it is written.

storage: the use of data structures to temporarily or permanently store data. Common data structures are variables, constants and arrays.

StreamReader: a class library within Visual Basic that provides access to code that is used to create text files and save data to text files.

string index: an index location given to each character within a string. The index is zero based from the first character.

structure diagrams: a method of expressing a system as a series of subsystems using a diagram.

subroutine: subroutines provide an independent section of code that can be called from another routine while the program is running. In this way, subroutines can be used to perform common tasks within a program.

substringing: a programming method used to identify part of a string value by reference to the index locations within that string.

syntax errors: mistakes made in the code equivalent to spelling and punctuation mistakes in English.

text file: a file that stores data as text. It normally has the file extension .txt.

top-down design: a way of designing a computer program by breaking down the problem into smaller problems (subsystems) until it is sufficiently defined to allow it to be understood and programmed.

trace tables: a way to test and find bugs in programs. A table is constructed to keep track of the values held in variables as a program is stepped through line by line.

two-dimensional array: an array with two index sets. Contains multiple rows of data with multiple columns. Each individual element identified by a combination of both row and column index.

validation: the process of programming a system to automatically check that data satisfies a set of specified input criteria; for example, passwords must be longer than six characters.

variable: a memory location used to store a value; the value of the data can be changed during program execution.

verification: a process that confirms the integrity of data as it is input into the system or when it is transferred between different parts of a system; for example, a CAPTCHA image used to prove data is being entered by a human.

Visual Studio 2019 Community: a version of the Integrated Development Environment (IDE) produced by Microsoft used to create programs using the Visual Basic programming language.

WHILE loop: a type of iteration that will repeat a sequence of code while a set of criteria continues to be met. If the criteria are not met before the loop starts, the loop will not run and the code within it will not be executed.

Windows Forms application: a GUI provided by Visual Studio 2019 Community used to create programs using the Visual Basic programming language. Windows Forms application is an event-driven approach to programming where subroutines are linked to GUI objects.

writing data to a file: the process of saving data to a text file.

zero-based index: the index is numbered from zero. The first character in the string will be at index zero. The final index location will always be one less than the number of characters in the string.